T0229649

MICROSCOPY HANDBOOKS 35

Negative Staining and Cryoelectron Microscopy:
the thin film techniques

Royal Microscopical Society MICROSCOPY HANDBOOKS

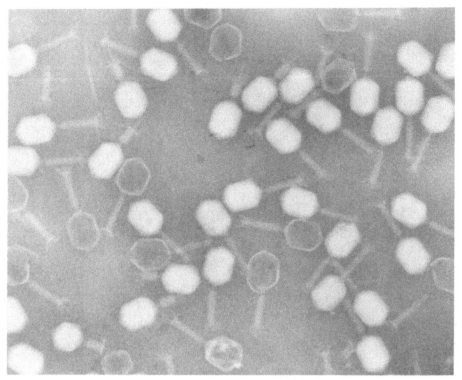

Frontispiece

Cryonegative staining of T2 bacteriophage, frozen directly from a thin film of 16% ammonium molybdate. Previously unpublished print, courtesy of Marc Adrian, Laboratory of Ultrastructural Analysis, University of Lausanne.

Negative Staining and Cryoelectron Microscopy:
the thin film techniques

J. Robin Harris
Institute of Zoology,
University of Mainz,
D-55099 Mainz,
Germany

Taylor & Francis
Taylor & Francis Group
LONDON AND NEW YORK

In association with the Royal Microscopical Society

© **Taylor & Francis Publishers Limited, 1997**

First published 1997

All rights reserved. No part of this book may be reproduced or transmitted, in any form or by any means, without permission.

A CIP catalogue record for this book is available from the British Library.

ISBN 1 85996 120 7

Published by Taylor & Francis
2 Park Square, Milton Park, Abingdon, Oxon, OX14 4RN
270 Madison Ave, New York NY 10016

Transferred to Digital Printing 2009

Typeset by Els Boonen, BIOS Scientific Publishers Ltd, Oxford, UK.

Front cover: Keyhole limpet haemocyanin didecamer crystallographic 2-D average, viewed from the side-on orientation. Colour print courtesy of Zdenka Cejka, Max-Planck-Institute for Biochemistry. Modified from Harris JR, Cejka Z, Wegener-Strake A, Gebauer W, Markl J. (1992) Two-dimensional crystallization, transmission electron microscopy and image processing of keyhole limpet haemocyanin. *Micron Microsc. Acta* 23, 287–301.

Publisher's Note
The publisher has gone to great lengths to ensure the quality of this reprint but points out that some imperfections in the original may be apparent.

Contents

Abbreviations

ACh	acetylcholine
AM	ammonium molybdate
CMC	critical micelle concentration
cpn	chaperonin
CTF	contrast transfer function
2-D	two-dimensional
3-D	three-dimensional
GON	group of nine (adenovirus hexons)
HIV	human immunodeficiency virus
KLH1/2	keyhole limpet haemocyanin types 1 and 2
10-MALT	n-decyl-β-D-maltopyranoside
MAP	microtubule associated protein(s)
M_r	relative molecular mass
MRC	Medical Research Council (UK)
MT	microtubule
ncd	non-claret disjunctional (protein)
NPC	nuclear pore complex
NS–CF	negative staining–carbon film (technique/procedure)
OG	n-octyl-β-D-glucopyranoside (octyl glucoside)
PBS	phosphate buffered saline
PEG	polyethylene glycol
pI	isoelectric point
8-POE	octylpolyoxyethylene
SEM	scanning electron microscope/microscopy
SFV	Semliki Forest virus
SIV	simian immunodeficiency virus
S-layer	surface layer (bacterial)
SLO	streptolysin O
STEM	scanning transmission electron microscope/microscopy
TEM	transmission electron microscope/microscopy
TMV	tobacco mosaic virus

Abbreviations

Preface

To write an EM techniques book on negative staining has long been in my mind. To actually get around to doing it was a different matter altogether! This was particularly so when it became clear that the more recently established thin film cryotechnique for the production of unstained specimens in vitreous ice should also be covered. Many helpful comments were received at the early planning and discussion stages, which were a prime force in encouraging me to proceed.

I have tried to produce a reasonably advanced and thorough survey of the available techniques for negative staining and for cryoelectron microscopy. Since much of the material currently available in reviews and elsewhere is now somewhat dated Chapters 5 and 8 cover a large number of biological applications from both unpublished and recently published studies by myself and others. Hopefully this approach, although it makes the present handbook longer than others in the series, provides direct access to the relevant literature and removes the need for its users to purchase another book. I hope that the book will be of interest and value to a broad range of biological electron microscopists (scientists, graduate students and technicians) and perhaps also to those microscopists with a more general interest in what is happening in present-day macromolecular electron microscopy.

An element of the *personal approach* has inevitably crept in; others may do things in a slightly different way, but the underlying principles will probably be the same. My view will usually be that the simplest techniques are often the best.

Robin Harris

Acknowledgements

I would like to gratefully acknowledge the encouragement and considerable assistance so freely provided by Jacques Dubochet and Marc Adrian, without whose help this booklet could not have been compiled. Much valuable help and comment on the negative staining section came from Milan Nermut and useful discussion from Prakash Dube. Thanks also go to the many scientists who so readily provided copies of figures from their published works, and in some instances copies of unpublished data. My selection of applications is inevitably limited and incomplete; so, I wish to emphasize the existence of a large volume of excellent published data from other authors that is not included or even mentioned here. The enthusiastic support provided by Bob Horne, over a period of several years, cannot be over-stated. Helen Saibil and Shaoxia Chen introduced me directly to the potential of cryoelectron microscopy, for which I remain very grateful. The more local support from my colleagues in the Institute of Zoology at the University of Mainz, in particular from Jürgen Markl and Albrecht Fischer, has been a source of strength and encouragement throughout the preparation of the manuscript. Elizabeth Sehn (EM Unit) provided useful technical comments during the early, difficult stages. Samples of bacterial toxin, used for several of the previously unpublished negative staining applications, were kindly provided by Sucharit Bhakdi, Institute of Medical Microbiology, University of Mainz. Finally, I would like to acknowledge the advice and encouragement I have received from Jonathan Ray, of BIOS Scientific Publishers Ltd, throughout the whole period of planning and writing of this book.

Safety

Attention to safety aspects is an integral part of all laboratory procedures, and both the Health and Safety at Work Act and the COSHH regulations impose legal requirements on those persons planning or carrying out such procedures.

In this and other Handbooks every effort has been made to ensure that the recipes, formulae and practical procedures are accurate and safe. However, it remains the responsibility of the reader to ensure that the procedures which are followed are carried out in a safe manner and that all necessary COSHH requirements have been looked up and implemented. Any specific safety instructions relating to items of laboratory equipment must also be followed.

1 Introduction

Since the commencement of biological studies using the transmission electron microscope (the TEM) in the late 1940s and early 1950s, scientists have sought to develop procedures for the preparation and investigation of thinly spread specimens of biological particulates. The word *particulates* will be used throughout in an extremely broad manner, to mean almost any finely dispersed biological material that can exist in an aqueous suspension. That is, material suspended in distilled water, low and increasing concentrations of buffer, through to a high salt or organic solute (e.g. sucrose, glycerol, urea) solution. Where appropriate, individual biological particulates will be named and discussed within the context of specimen preparations and electron microscopical applications. Thus, I refer here primarily to biological material in the form of isolated subcellular membranes and components such as ribosomes and nuclear pore complexes, nucleosomes, protein molecules (soluble and fibrous) and their higher oligomeric assemblies or complexes, bacteria and their appendages, viruses, liposomal vesicles and reconstituted membrane systems.

The metal shadowing technique for thinly spread biological particles was introduced well ahead of negative staining. High resolution single angle and rotary metal shadowing, for instance with platinum–carbon, tantalum–tungsten and chromium metal shadowing, continues to be an important method for contrasting biological particles. In particular, extended molecules such as nucleic acids and elongated/filamentous proteins have been successfully studied. Metal shadowing will not be dealt with further and the reader is referred to useful surveys by Gross (1987), Nermut and Eason (1989) and Slayter (1981, 1991) and the recent high resolution scanning electron microscopy (SEM) study on reoviruses from Centonze *et al.* (1995).

Negative staining was established in its own right as a specimen preparation technique by the late 1950s, primarily due to the efforts of Robert (Bob) Horne and his colleagues, when working at the Institute of Animal Physiology, Babraham, Cambridge (as outlined more fully in Chapter 2). The techniques of negative staining will be presented in some detail (Chapters 3, 4 and 5) as will the more recent technical innovation of thin frozen-hydrated/vitrified unstained specimen preparation and cryotransmission electron microscopy (Chapters 6, 7 and 8).

1

Negative staining became widely used throughout the 1960s and 1970s, particularly in the hands of microbiologists and virologists such as David Gregory, Milan Nermut, Herman Frank, Roy Markham, Peter Wildy and Bob Horne, for both diagnostic and research applications on bacteria and a wide range of animal, human, plant and bacterial viruses (for reviews see Horne, 1991; Horne and Wildy, 1979). The early molecular structural biologists, particularly Aaron Klug and his colleagues at the MRC Laboratory for Molecular Biology in Cambridge, rapidly made a major contribution. Negative staining was also taken up by a number of biochemists and cell biologists, notably Ed Munn in Babraham, Ennio Lucio Benedetti in Amsterdam, Erni van Bruggen in Groningen, the Netherlands, and Rudy Haschemeyer, Robert Oliver and Robley Williams in the USA. Erni van Bruggen established important negative staining studies on the haemocyanins from several invertebrate species, work that has continued actively through to the present day in the Groningen group and several others throughout the world, including those of Jean Lamy in Tours, Marin van Heel in Berlin and Jürgen Markl in Mainz.

Successful cryoelectron microscopy of unstained vitrified specimens followed on from earlier ideas on specimen freezing by Humberto Fernández-Morán, Bob Glaeser and Ken Taylor. It was developed as a standard preparation technique for biological materials almost entirely due to the persistent efforts of Jacques Dubochet and his colleagues, in particular Marc Adrian and Jean Lepault, at the European Molecular Biology Laboratory in Heidelberg in the late 1970s and early 1980s. It should perhaps be emphasized that these technical innovations at the specimen level were accompanied by TEM instrumental improvements that enabled specimens to be successfully studied at low temperatures (i.e. specimen cryoholders, cryostages, efficient anti-contaminators and low electron dose systems). In the following decade cryoelectron microscopy of thin unstained vitrified specimens was utilized by many researchers, often in parallel with the simpler and more rapid negative staining approach, for numerous biological applications, including extensive studies on viruses (Baker *et al.*, 1991; Chiu, 1993). Although many initially thought that negative staining would by now have been completely overtaken and overshadowed by cryoelectron microscopy, in reality this has not happened. The two technical approaches continue in parallel, although it can be acknowledged that the continued use of negative staining by some investigators, rather than vitrification, might often be determined by the expense and therefore still restricted availability of cryo-TEMs and the greater technical difficulty of the cryopreparation techniques. Furthermore it has been indicated by some (Bremner *et al.*, 1992) that the gains from cryo-TEM of unstained vitrified specimens may be marginal in terms of improved resolution. From theoretical considerations of protein electron density, negative staining is thought only to be able to resolve the surface shell of biological structures, whereas cryoelectron microscopy of unstained biological material has the potential also to reveal internal structural information, if a sufficient number of particles can be averaged for this information to be retrieved at high resolution.

For viral three-dimensional (3-D) reconstructions from vitreous ice the best resolution achieved is often in the order of 25 Å. For the bacterial ribosome Frank *et al.* (1995) have achieved 24 Å[a]. For the molluscan haemocyanin didecamer the single particle 3-D resolution from ice has been in the order of 40–45 Å (Dube *et al.*, 1995a; Lambert *et al.*, 1995) whereas from low temperature ammonium molybdate–glucose studies a figure of 15 Å has recently been obtained (Dube *et al.*, 1995b; Orlova *et al.*, unpublished obeservations). Nevertheless, with two-dimensional (2-D) crystalline specimens, such as purple membrane bacteriorhodopsin, bacterial porins and plasma membrane ion channels, the obtainable resolution from low temperature studies in the presence of glucose or trehalose has been considerably superior for both 2-D and 3-D reconstructions, although the recovery of information in the third dimension is always somewhat inferior to that in the planar dimension. For the acetylcholine receptor, in the form of helical tubes, studies in vitreous ice have produced a 3-D reconstruction at 10 Å, indicating that the future study of unstained 2-D crystals in vitreous ice may hold considerable potential for further improvement of resolution.

It is the intention of this book to present, in a somewhat personal manner, the possibilities and limitations of both negative staining and cryoelectron microscopy of unstained vitreous specimens. A comparative assessment will be given in a balanced manner, with inclusion of many biological applications from both approaches. Although the negative staining technique is by far the older of the two, and has been exploited in many areas for many years, there is still considerable possibility for further development (Harris and Horne, 1994), particularly in the area of low temperature with the inclusion of trehalose or other protective agents (Harris *et al.*, 1995). As a relatively new technique, the cryoelectron microscopy of thin vitrified unstained specimens is still undergoing a period of considerable expansion, which is likely to continue into the foreseeable future. The combination of negative staining with vitrification, which for lower resolution studies should offer increased contrast and benefits of cryoprotection, has yet to be seriously investigated (see Chapter 9). There are indications that the presence of trehalose in aqueous unstained samples prior to vitrification may also provide additional protection of biological samples, both during rapid freezing and subsequent electron irradiation (Marc Adrian, personal communication).

---

[a] Because it is conventional in X-ray crystallography to use Ångstrom (Å) units rather than nanometers (nm) for molecular dimensions, together with instrumental and molecular resolutions, electron microscopists dealing with macromolecules and viruses often tend to follow the same convention. Scale bars on original electron micrographs will, however, usually be given in nm.

References

Baker TS, Newcombe WW, Olson NH, Cowsert LM, Olson C, Brown JDC. (1991) Structure of bovine and human papilloma viruses–analysis by cryoelectron microscopy and 3-D image-reconstruction. *Biophys. J.* **60**, 1445–1456.

Bremner A, Henn C, Engel A, Baumeister W, Aebi A. (1992) Has negative staining a future in biomolecular electron microscopy? *Ultramicroscopy* **46**, 85–111.

Centonze VE, Chen Y, Severson TF, Borisy GG, Nibert ML. (1995) Visualization of individual reovirus particles by low-temperature high-resolution scanning electron microscopy. *J. Struct. Biol.* **115**, 215–225.

Chiu W. (1993) What does electron cryomicroscopy provide that X-ray crystallography and NMR spectroscopy cannot? *Ann. Rev. Biophys. Biomol. Str.* **22**, 233–255.

Dube P, Orlova EV, Zemlin F, van Heel M, Harris JR, Markl J. (1995a) Three-dimensional structure of keyhole limpet hemocyanin by cryoelectron microscopy and angular reconstitution. *J. Struct. Biol.* **115**, 226–232.

Dube P, Stark H, Orlova EV, Schatz M, Beckmann E, Zemlin F, van Heel M. (1995b) 3D structure of single macromolecules at 15 Å resolution by cryo-microscopy and angular reconstitution. In *JMSA Proceedings, Microscopy and Microanalysis* (eds GW Bailey, MH Ellisman, RA Hennigar, NJ Zaluzec). Jones and Begell, New York, pp. 838–839.

Frank J, Zhu J, Penczek P, Yanhong L, Srivastava S, Verschoor A, Radermacher M, Grassucci R, Lata R, Agrawal RK. (1995) A model of protein synthesis based on cryo-electron microscopy of the *E. coli* ribosome. *Nature* **376**, 441–444.

Gross H. (1987) High resolution metal replication of freeze-dried specimens. In *Cryotechniques in Biological Electron Microscopy* (eds RA Steinbrecht, K Zierold). Springer-Verlag, Berlin and Heidelberg, pp. 205–215.

Harris JR, Horne RW. (1994) Negative staining: a brief assessment of current technical benefits, limitations and future possibilities. *Micron* **26**, 5–13.

Harris JR, Gebauer W, Markl J. (1995) Keyhole limpet hemocyanin: negative staining in the presence of trehalose. *Micron* **26**, 25–33.

Horne RW. (1991) Early developments in the negative staining technique for electron microscopy. *Micron Microsc. Acta* **22**, 321–326.

Horne RW, Wildy P. (1979) An historical account of the development and applications of the negative staining technique to the electron microscopy of viruses. *J. Microsc.* **117**, 103–122

Lambert O, Traveau, J-C, Boisset N, Lamy JN. (1995) Three-dimensional reconstruction of the hemocyanin of the protobranch bivalve mollusc *Nucula hanleyi* from frozen-hydrated specimens. *Arch. Biochem. Biophys.* **319**, 231–243.

Nermut MV, Eason P. (1989) Cryotechniques in macromolecular research. *Scann. Microsc.* Suppl. 3, 213–225.

Slayter H. (1981) Electron microscopy of glycoproteins. In *Electron Microscopy of Proteins* (ed. JR Harris) Vol. 1. Academic Press, London, pp. 197–254.

Slayter H. (1991) High resolution shadowing. In *Electron Microscopy in Biology: a Practical Approach* (ed. JR Harris). IRL Press, Oxford, pp. 151–172.

2 Negative Staining: Historical Background and Technical Development

There is a considerable body of information of a technical and review nature on negative staining elsewhere in the literature (Bremner *et al.*, 1992; Harris and Horne, 1991, 1994; Hayat and Miller, 1990; Holzenburg, 1988; Horne, 1991; Nermut, 1991; Spiess *et al.*, 1987; Valentine and Horne, 1962); therefore, throughout this handbook I shall attempt to avoid undue repetition and present the current laboratory approaches with which I am personally familiar and have achieved some technical success. After considering the range of procedures necessary for the production of carbon and carbon–plastic support films (sometimes termed substrates) emphasis will be placed upon the *droplet* negative staining technique (Harris and Agutter, 1970; Harris and Horne, 1991). I was introduced to this approach by Ennio Lucio Benedetti in his laboratory, at the start of my doctoral studies in 1966 (Anderson, 1966; Benedetti and Emmelot, 1965, 1968) and with minor variations it is still useful for many biological applications. Secondary to this approach, but with perhaps considerably greater scope for further technical development, are the *floating method* and the *negative staining–carbon film* (NS–CF) procedures. Actually, both of these are 'floating' variants of the negative staining technique and have their roots in the mica–carbon film production procedure (Section 3.1), using the inherent property that a freshly released floating layer of carbon has a strong adsorptive and stabilization capacity for proteins and viruses (Valentine *et al.*, 1966). When developed further by Ivonne Pasquali-Ronchetti and Bob Horne in the early 1970s by spreading virus suspensions in the presence of ammonium molybdate on a mica surface *before* carbon coating, this led to the production of ordered arrays of viruses attached to the carbon film, which subsequently could be negatively stained (for details see Section 3.3); thus the proposal of the rather verbose terminology 'negative staining–carbon film technique'. My own interest in the NS–CF procedure for the production of ordered arrays and 2-D crystals of protein molecules commenced soon after the initial description of this approach (Horne and Pasquali-Ronchetti, 1974) and continues strongly to this day (Harris, 1991; Harris and Holzenburg, 1989, 1995).

The concept of negative staining as a light microscopical procedure that will enable an essentially transparent object to be rendered visible by surrounding it with a coloured solution is rather old, but the transfer of this idea to thinly spread material for electron microscopy came only slowly. It

5

is widely accepted that Bob Horne was the first electron microscopist to really come to terms with the subject. He presented clear data showing that bacteriophages could be surrounded by a thin amorphous layer of air-dried sodium phosphotungstate or uranyl acetate, which considerably reduced the structural flattening that occurred in the absence of stain (Brenner and Horne, 1959). Early attempts to positively stain protein molecules and viruses with low pH phosphotungstate also gave some indication of negative staining when the stain was in excess or incompletely washed away, but this was not established independently as a routine procedure (Hall, 1955; Huxley and Zubay, 1960). Clearly, in this early developmental work it was important to select heavy metal salts that air-dried to give an evenly spread non-crystalline amorphous or vitreous glass-like electron-opaque or -dense layer surrounding and supporting (i.e. embedding) the biological particles.

Negative staining is believed to reduce the surface tension forces at the air–fluid interface that would otherwise produce considerable flattening of biological material on to a support film and the actual thickness of the dried stain layer is also likely to be rather important. The thickness of the dried negative stain will be determined by the concentration and the volume of fluid dried on to the support film. As indicated below (also Section 3.2) the thickness of the supporting layer of dried stain can be considerably increased by including a carbohydrate such as glucose or trehalose, which is also likely to protect protein structure during specimen drying. Ideally, negative stains should not interact with biological material in a 'positive' staining manner, neither must they cause any aggregation or precipitation, or indeed molecular dissociation (but see Sections 3.3.2 and 3.5.6). A very limited number of heavy metal salts were found to be suitable as negative stains, and in all cases it was necessary to avoid the additional presence of sodium chloride, buffer salts and routine biochemical reagents such as urea, glycerol and sucrose, other than as trace quantities.

The most useful negative staining salts defined by the early investigators are uranyl acetate, uranyl formate, sodium/potassium phosphotungstate, sodium silicotungstate and ammonium molybdate. Other, lesser used salts will be mentioned below (Section 3.2.3). That these salts all have the property of drying as an amorphous layer does not necessarily mean that microcrystallites of stain are not present in this layer. Indeed, this situation and the possible lability and even mobility of such crystallites within the electron beam has been addressed (Unwin, 1974). Since the ability of any negative stain to reveal structural detail must depend upon permeation of the staining cation or anion into aqueous cavities inside a biological particle, this could be limited by the size of such crystallites, but perhaps ultimately by the hydrated ionic diameter of the negative staining salt *in solution*. Thus, it is likely that negative staining will reveal primarily the outer surface and therefore the overall shape/replica or quaternary structure of a protein molecule, and be unable to reveal high resolution detail (i.e. better than c. 15 Å; but see Kiselev *et al.*, 1990, and Chapter 9). Certainly, the presence of surface regions containing a number of hydrophobic amino acids and hydrophobic pockets within a protein are unlikely to be optimally negatively stained. Regions containing a net positive or net negative charge are likely to interact differently with anionic and cationic negative stains, respectively. For instance, the uranyl cation is known to bind to phosphate groups and also to carboxyl groups, whereas the molybdate anion is likely to sequester bound calcium and magnesium and may also interact directly with available positively charged amino groups. These more detailed considerations apart, the negative stains generally appear to behave in a similar manner, based upon the overall comparability of the electron optical images in different stains. Nevertheless, significant staining differences may occur (Woodcock and Baumeister, 1990). During the several decades since the establishment of the negative staining procedure numerous scientists have successfully utilized this approach for the study of a wide range of biological material, resulting in a large literature. Due credit should be given to those working within the group of Wolfgang Baumeister (in Martinsried), those in the group of Erni van Bruggen (in Groningen) those with Joachim Frank (in New York) and José Carrascosa (in Madrid), and those in the groups of Ueli Aebi and Andreas Engel (in Basel). Of the many individuals who have made a significant contribution over the years, the names of Linda Amos, Egbert Boekema, Roger Craig, Howard Egelman, Andeas Holzenburg, Kevin Leonard, Bill Massover, Milan Nermut and Alasdair Steven might be mentioned.

After a lengthy period of procedural stability and lack of progress, I feel that further technical development and a more critical assessment of the potential of negative staining is now required (Harris and Horne, 1994). The routine inclusion of a carbohydrate such as glucose or trehalose (Harris *et al.*, 1995) along with negative stain, combined with low temperature TEM study, appears to provide some improvement of imaging. At the same time this carbohydrate–stain mixture creates a thicker amorphous layer within which protein molecules can more freely adopt varying orientations with respect to the carbon support film, with reduced adsorptive interaction and the likelihood of structural changes or restriction of stain access because of close apposition of protein and carbon. This thicker layer of stain and trehalose considerably reduces the undesirable flattening of tubular structures or other fragile structures such as liposomes (see Section 3.2 and Chapter 5). Trehalose has the unique ability to create vitreous/glass-like carbohydrate–water films in which the water is tightly bound (Green and Angell, 1989), which may explain its ubiquitous and extremely useful property as a biological protectant at extremes of temperature and dehydration, and to UV irradiation. Study of negatively stained specimens at low temperature in a specimen holder cooled by liquid nitrogen or liquid helium, although not new, has until recently (Harris *et al.*, 1995) been used primarily by those investigating thin 3-D and 2-D protein crystals.

In the presence of negative stain and trehalose, or glucose, low temperature ($-170°C$, or below) and low electron dose study is certainly desirable to obtain the best resolution from biological material (Kiselev *et al.*, 1990). Nevertheless, with trehalose, room temperature studies can be performed providing the stain–carbohydrate layer is relatively thin and a low electron doses maintained, otherwise pronounced 'bubbling' can occur (see Section 4.3). Although in the early work in this area only combinations of uranyl acetate and glucose were used, it may be significant that more recent studies have also utilized ammonium molybdate together with glucose or trehalose. From theoretical considerations ammonium molybdate is likely to be a superior negative stain to uranyl acetate, since it is not generally known to impart any element of positive staining. It is likely that this approach can be extended to combinations of trehalose with any of the known anionic negative staining salts.

References

Anderson TF. (1966) Electron microscopy of microorganisms. In *Physical Techniques in Biological Research* (ed. AW Pollister) Vol. III, Part A. Academic Press, New York, pp. 319–387.

Benedetti EL, Emmelot P. (1965) Electron microscopic observations on negative staining of plasma membrane isolated from rat liver. *J. Cell Biol.* **26**, 299–305.

Benedetti EL, Emmelot P. (1968) Hexagonal array of subunits in tight junctions separated from rat liver plasma membranes. *J. Cell Biol.* **38**, 15–24.

Bremner A, Henn C, Engel A, Baumeister W, Aebi A. (1992) Has negative staining a future in biomolecular electron microscopy? *Ultramicroscopy* **46**, 85–111.

Brenner S, Horne RW. (1959) A negative staining method for high resolution electron microscopy of viruses. *Biochim. Biophys. Acta* **34**, 60–71.

Green JL, Angell CA. (1989) Phase relations and vitrification in saccharide-water solutions and the trehalose anomaly. *J. Phys. Chem.* **93**, 2880–2882.

Hall CE. (1955) Electron densitometry of stained virus particles. *J. Biophys. Biochem. Cytol.* **1**, 1–12.

Harris JR. (1991) The negative staining–carbon film procedure: technical considerations and a survey of macromolecular applications. *Micron Microsc. Acta* **22**, 341–359.

Harris JR, Agutter PS. (1970) A negative staining study of human erythrocyte ghosts and rat liver nuclear membranes. *J. Ultrastruct. Res.* **32**, 405–416.

Harris JR, Holzenburg A. (1989) Transmission electron microscopic studies on the quaternary structure of human erythrocyte catalase. *Micron Microsc. Acta* **20**, 223–238.

Harris JR, Holzenburg A. (1995) Human erythrocyte catalase: 2-D crystal nucleation and production of multiple crystal froms. *J. Struct. Biol.* **115**, 102–112.

Harris JR, Horne RW. (1991) Negative staining. In *Electron Microscopy in Biology: a Practical Approach* (ed. JR Harris). IRL Press, Oxford, pp. 203–228.

Harris JR, Horne RW. (1994) Negative staining: a brief assessment of current technical benefits, limitations and future prospects. *Micron* **25**, 5–13.

Harris, JR, Gebauer W, Markl J. (1995) Keyhole limpet haemocyanin: negative staining in the presence of trehalose. *Micron* **26**, 25–33.

Hayat MA, Miller SM. (1990) *Negative Staining*. McGraw-Hill, New York.

Holzenburg A. (1988) Preparation of 2-D arrays of soluble proteins as demonstrated for bacterial D-ribose-1,5-bisphosphate carboxylase/oxygenase. *Meth. Microbiol.* **20**, 341–356.

Horne RW. (1991) Early developments in the negative staining technique for electron microscopy. *Micron Microsc. Acta* **22**, 321–326.

Horne RW, Pasquali-Ronchetti I. (1974) A negative staining–carbon film technique for studying viruses in the electron microscope. I. Preparative procedure for examining icosahedral and filamentous viruses. *J. Ultrastruct. Res.* **47**, 361–383.

Huxley HE, Zubay G. (1960) Electron microscope observations on the structure of microsomal particles. *J. Mol. Biol.* **2**, 10–18.

Kiselev NA, Sherman MB, Tsuprun VC. (1990) Negative staining of proteins. *Electr. Microsc. Rev.* **3**, 43–72.

Nermut MV. (1991) Unorthodox methods of negative staining. *Micron Microsc. Acta* **22**, 327–339.

Spiess E, Zimmermann H-P, Lunsdorf H. (1987) Negative staining of protein molecules and filaments. In *Electron Microscopy in Molecular Biology: a Practical Approach* (eds J Sommerville, U Scheer). IRL Press, Oxford, pp. 147–166.

Unwin PNT. (1974) Electron microscopy of the stacked disk aggregate of TMV protein. II. The influence of electron irradiation on the stain distribution. *J. Mol. Biol.* **87**, 657–670.

Valentine R, Horne RW. (1962) An assessment of negative staining techniques for revealing ultrastructure. In *The Interpretation of Ultrastructure* (ed. RHJ Harris). Academic Press, New York, pp. 283–278.

Valentine RC, Wrigley NG, Scrutton MC, Irias JJ, Utter MF. (1966) Pyruvate carboxylase: VIII. The subunit structure as examined by electron microscopy. *Biochemistry* **5**, 3111–3116.

Woodcock CL, Baumeister W. (1990) Different representation of protein structure obtained with different negative stains. *Eur. J. Biochem.* **51**, 45–52.

3 Negative Staining: Preparative Procedures

Since the early 1960s, much has been written regarding the preparation of thin plastic, plastic–carbon and carbon films, their varying properties and the treatments given to them before use as specimen supports for negative staining and other purposes (Hayat and Miller, 1990; Sommerville and Scheer, 1987). Often, one has the impression that electron microscopists, similar to all other scientists, revel in the reinvention or improvement of the wheel. Sometimes there are technical improvements and something really new, but often there are only minor innovations, with apparent oblivion to, or careless neglect of, earlier technical publications! To quote Thomas F. Anderson (1966) "Unfortunately, the original literature on techniques is becoming so voluminous that it is impossible for one to keep himself informed of all current advances", so it is hardly surprising that 30 years later the situation is even worse. In reality, the various negative staining procedures given below tend to be variants of a very limited number of 'original themes'. This inherent limitation apart, it is my intention to present some of the principal approaches, when possible giving a number of alternatives, with emphasis upon the overall simplicity of the techniques. Details of the three main approaches for the production of negatively stained specimens will be given below (Sections 3.3, 3.4, 3.5); namely, the droplet, floating and mica–carbon transfer techniques. Whilst at first glance it might appear that these techniques are totally independent of one another, this is not the case. There is considerable interplay between the techniques and indeed a close relationship to the techniques used for the preparation of support films. Again, this indicates the overall sparsity of technical possibilities, which has led investigators to seek out the most useful combinations. It will be noted by some that the spraying techniques are not included. In a safety-conscious scientific world, I think that these techniques should not be routinely employed unless absolutely necessary, and then only performed within an appropriate negative pressure extraction cabinet (see also Section 3.3). In general, suitable precautions should always be observed when handling viruses and potentially toxic or dangerous samples, with the inclusion of UV or chemically produced inactivation when necessary.

3.1 Carbon support films

3.1.1 Preparation of carbon films on mica

Given the availability of a carbon coating unit (the one essential but expensive piece of equipment that is necessary for the preparation of carbon support films required for making negatively stained specimens) the creation of a carbon layer on the clean surface of freshly cleaved mica is one of the oldest approaches. Pieces of mica, of appropriate size (2.5 cm × 7.5 cm) are cleaved by carefully inserting a needle or the points of fine forceps. The mica is then placed with its clean (untouched) inner side surface uppermost on a small sheet of white paper, under a carbon source within a coating unit. The carbon can be evaporated from a heated carbon rod, a carbon filament or a carbon electron beam source (in general, follow the manufacturer's instructions). An abbreviated procedure for the use of a carbon rod source in the Edwards Model 306 vacuum coating unit is given below: *Figure 3.1* shows the detail of a carbon rod source, set for evaporation, with the shield removed and in position above two pieces of mica, within a coating apparatus.

Procedure. When a vacuum of at least 10^{-4} Torr, preferably 10^{-5} Torr (1 Torr = 1.3 mbar = 1.3 hPa)[a] has been reached within the coating unit, carbon should be evaporated from the low voltage supply (10–30 V), by rapid and brief repeated maximal increase of the current control until the desired thickness of carbon has been deposited on the mica surface. (Usually four or five rapid 'firings' of the carbon will be sufficient.) This can readily be determined by eye using the density of the light grey carbon deposited on the nearby paper, or with a calibrated quartz crystal thickness monitor, to a thickness of approximately 10–40 nm (only possible with a single continuous 'firing'). With a little experience the former approach can be perfectly adequate, and carbon films of varying thickness can be produced for different purposes. In general, the thinner the carbon the more fragile it will be. Therefore, for high molecular weight fibrillar material or indeed cellular material, which is to be subjected to several on-grid washes, the use of slightly thicker carbon will probably be advantageous. Some workers think that carbon evaporation at a high vacuum (e.g. 10^{-6} Torr) will improve the strength of extremely thin carbon films. Nevertheless, if reasonable care is taken when handling the grids it is unlikely that all the carbon spanning the grid squares will break during a negative staining procedure (see Section 3.3.1) and the use of a platinum bacteriological loop for handling the grids through a series of droplet or spot-plate washings should help to preserve the integrity of even very

---

[a] 1 hPa (100 Pa) = 1 N m^{-2} = 1 mBar; 1 Pa = 1 N m^{-2} (= 1 m^{-1} kg sec^{-2}); 1 Torr = 101325/760 Pa = 1.333224 × 10^{2} Pa)

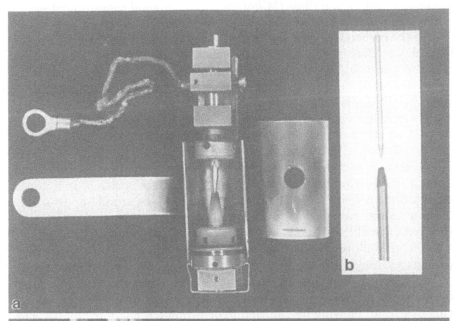

Figure 3.1: Carbon rod source.
(a) A carbon rod source, with the shield removed to show the two rods, pressed together by a metal spring. (b) Carbon rods shown separately and (c) the carbon rod source (S) in position within a coating unit for in vacuum evaporation of carbon on to two pieces of mica (M).

thin (e.g. 5 nm) carbon films (Milan Nermut, personal communication).

The mica + carbon layer should be left overnight before attempting to remove the carbon film by floating it on to a water surface. The reason for this delay is that freshly deposited carbon may not readily be released from the mica; this can be speeded up (not always reliably) by placing freshly prepared mica–carbon on a moist filter paper inside a petri dish and/or briefly applying moist breath on to the carbon surface. The film of carbon can be conveniently released on to the surface of distilled water in a 9 cm diameter petri dish or a similarly commercially available shallow glass apparatus with a central hole and outflow control. With the petri dish technique, individual bare grids (dull-side up) can be brought from beneath the floating carbon film to rapidly create a series of individual support films (*Figure 3.2*). The disadvantage of this simple approach is that often some carbon will fold around the edge of the grid, producing areas with a double-carbon layer, despite careful wiping with filter paper. In the other 'bulk production' approach (*Figure 3.3*) a series of bare grids, dull-side up, are positioned under distilled water on a strip of filter paper (e.g. Whatman No. 1, smooth surface up) held in a small stainless steel wire-gauze cage. The floating film of carbon is positioned over the array of grids by careful blowing by mouth and/or through a pasteur pipette. The water is then allowed to drain slowly out, allowing the carbon film to settle gently on to the grids. The stainless steel cage is then lifted out and dried under the gentle heat from an angle lamp, before removing the filter

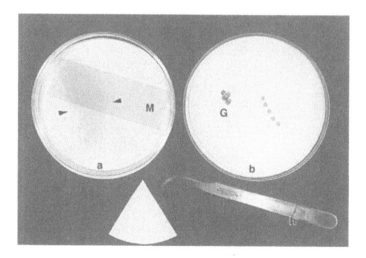

Figure 3.2: The production of single carbon support films in a 9 cm plastic petri dish containing distilled water.

On the water surface a floating layer of carbon (arrowed, a) has been released directly from a piece of carbon-coated mica (M) (produced as in *Figure 3.1*). Individual bare grids (G) (e.g. 300 or 400 mesh) are brought individually from beneath the floating carbon (dull/matt surface uppermost), carefully wiped on a filter paper to remove excess carbon, folded around the grid edge and the carbon film allowed to dry (b) [(h) Sliding holding ring of silicone rubber tubing.]

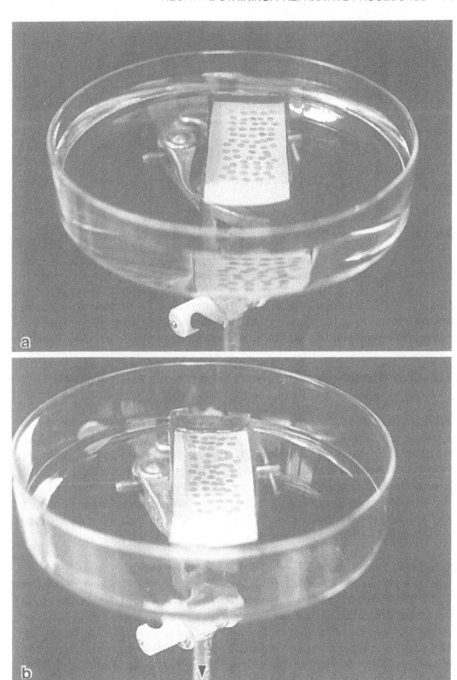

Figure 3.3: A shallow glass apparatus for the bulk production of carbon support films.
    (a) Under distilled water, bare 300 or 400 mesh grids are positioned on a piece of filter paper inside a stainless steel gauze, dull/matt surface upermost. A floating film of carbon (see *Figure 3.2*) is then positioned carefully over the grids and the water allowed to flow out from the apparatus (b), allowing the carbon to settle gently on to the grids as the water flows out.

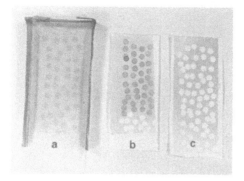

Figure 3.4: Carbon-coated grids on a filter paper within a stainless steel gauze cage.
(a) After drying and removal of the filter paper from the cage. (b) After removal of several grids. (c) After removal of all the grids for use. Note that in these examples the carbon has been made excessively thick, to demonstrate its presence photographically.

paper + grids and carbon support film (*Figure. 3.4*). (Note that the carbon films are more likely to break if removed immediately from the surface of the wet filter paper, due to surface tension forces.)

3.1.2 Preparation of carbon films on a thin plastic support

With this alternative procedure for the preparation of carbon support films it is necessary first of all to prepare a thin plastic support/substrate, made of formvar (polyvinyl formvar/formal), butvar (polyvinyl butyl), polyvar, parlodion, collodion (nitrocellulose) or pyroxylin. It has been claimed that the newer product, pioloform, has a greater mechanical and thermal stability than formvar or collodion and also a lower mass–thickness for a given thickness of plastic film. Cellulose acetate, cellulose triacetate (Triatol) and Bioden are also recommended (Agar Scientific Ltd).

Procedure. Prepare a 0.1% (w/v) solution of the polyvinyl plastic in acetone, chloroform, ethylene dichloride or amyl acetate (allow plenty of time for the solution to clarify completely, i.e. overnight). A cleaned glass microscope slide should then be dipped lengthwise (approx. 80%) into the plastic solution, withdrawn and drained on to a filter paper. After drying completely, a scalpel or razor blade should be scraped gently along the three edges of one surface of the slide and the thin layer of plastic floated on to the surface of distilled water in an appropriate container (*Figure 3.5*). Bare grids can then be placed dull-surface down on to the floating plastic, and the plastic film + grids removed by a single *confident* movement using a small piece of paper, (or Parafilm) with its ends slightly folded or a piece of Perspex with a small handle fixed centrally. (Alternatively, the plastic film + grids can be positioned over and lowered on to a

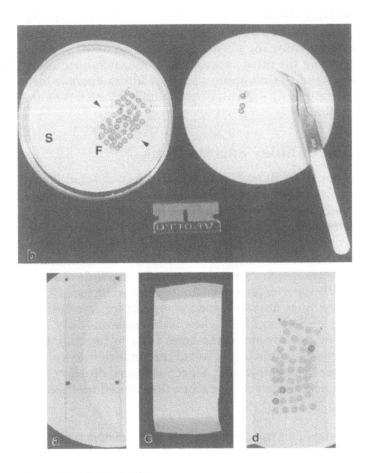

Figure 3.5: Formvar strips.
(a) A formvar-coated microscope slide (with the position of the formvar marked by four black dots). (b) A formvar strip (F) released from a microscope slide (S), floating on a water surface, with grids positioned dull/matt surface facing *down*, ready to be removed with a filter paper, Parafilm or Perspex strip (c), and after drying and ready for carbon-coating (d).

piece of cut filter paper in the stainless steel cage, as described in Section 3.1.1, above.) The paper (or Parafilm + grids) should then be coated with a thin layer of carbon (10–40 nm), as in Section 3.1.1. Grids prepared in this way can be used directly, but for most purposes the plastic should be removed by washing with chloroform or amyl acetate solvent. Note that if such grids are used directly they are likely be considerably more robust than carbon alone, but that they will have different spreading and adsorption properties, before and after glow-discharge treatment (see Section 3.1.4). Also note that with the somewhat thicker carbon–plastic support films, the electron beam intensity in the TEM will need to be slightly higher to provide optimal image brightness, thereby producing more rapid damage to the specimen material. Chloroform washing can be performed with *individual* grids, by inserting them vertically for 10 sec in a solution in a

50 ml beaker and repeating the procedure in a second beaker. (This is the *only* possible way to remove the plastic if the grids are on Parafilm or a Perspex support.) Alternatively, the plastic from a whole batch of grids can be dissolved by placing the filter paper strip (alone or in a stainless steel cage) in a glass petri dish containing a stack of filter papers, which are then carefully saturated with solvent, covered and left for a few hours in a fume-extraction cabinet.

3.1.3   Preparation of holey carbon films

Holey carbon films are carbon films containing an array of small holes, which may be of a regular size or somewhat variable, depending upon the procedure used for their preparation. Such films have also been termed *carbon nets, perforated carbons, fenestrated carbons* or *carbon micro-grids*, to mention a few alternative names (Fukami and Adachi, 1965; Jahn, 1995). Holey carbon films can be used for negative staining, with the intention of studying biological materials such as membranes or 2-D crystals that are spanning the holes and supported by the surrounding carbon (see Chapter 9). Such holey carbon films are the standard support for cryoelectron microscopy specimens (see Section 7.1.1), where a thin aqueous layer containing the material under study spans the holes prior to freezing. It is sometimes useful to use a *very* thin (e.g. 3 nm of carbon layer) across the holes in a slightly thicker holey carbon film, for conventional droplet negative staining or for the negative staining–carbon film procedure (Section 3.5).

Procedure. Two procedures for the preparation of holey carbon films will be given, both of which are related to the plastic film technique given above. Firstly, prepare a thin plastic film, as in Section 3.1.2, to the stage when the glass microscope slide is removed from the organic solvent–plastic solution and rapidly drained on to a filter paper; however, *before* the plastic dries, repeatedly breathe on to the drying plastic on one surface of the microscope slide. The small water droplets in the moist breath will perforate the plastic just at the position where the drying zone is visible, leaving a series of slightly opaque arcs across the microscope slide. The holes produced by this procedure are somewhat variable in size and number, but are usually perfectly satisfactory. The perforated plastic should then be floated on to water, bare grids placed upon it, and carbon coated, as in Sections 3.1.1 and 3.1.2. above. Again, the plastic is usually dissolved in organic solvent before the grids are used. This is necessary to convert the plastic *pseudoholes* into true holes (see *Figure 3.6*).

   An alternative procedure for the preparation of holey films uses an ultrasonicated or vigorously hand shaken emulsion of glycerol (0.1% v/v), or a 1:5 (v/v) mixture of neutral detergent and glycerol, in the 0.1% (w/v) solution of formvar or pioloform in organic solvent. Microscope slides are dipped into the glycerol–plastic–solvent emulsion, as above (Section 3.1.2),

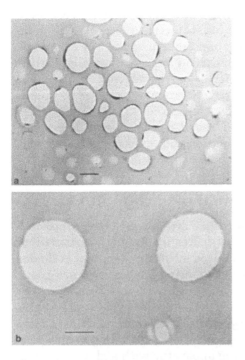

Figure 3.6: A holey carbon support film.
(a) A region of holes in a holey/perforated carbon support film produced by breathing on to drying formvar on a microscope slide, followed by release, carbon coating (as in *Figures 3.1* and *3.5*), and removal of the formvar by individually washing the grids in chloroform. Only a few small 'pseudoholes' remain. (b) A higher magnification region, showing two holes and the smooth carbon surface alongside. The scale bars indicate 250 nm and 100 nm, respectively.

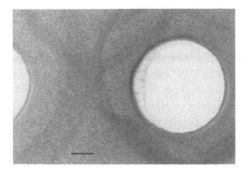

Figure 3.7: A holey carbon film coated with gold.
An example of part of a holey carbon film, with the carbon coated with gold particles (by sputter-coating), as a focusing aid for cryoelectron microscopy (courtesy of Marc Adrian). The scale bar indicates 200 nm.

drained and air-dried. The dried plastic film is then floated off the plastic, as above, bare grids are positioned and the plastic film + grids recovered on a piece of filter paper. The glycerol (or surfactant/glycerol) can be removed by placing the paper on a stack of methanol-soaked filter papers. After drying, the grids are coated with carbon *in vacuo*, usually slightly more thickly than would be the case for a continuous carbon film. This procedure can produce holes of an extremely regular size in the plastic and carbon film, the size being dependent upon the extent of emulsification (i.e. ultrasonication/shaking time); see also Jahn (1995). Holey carbon support films can also be sputter-coated with gold particles, as a focusing aid for cryoelectron microscopy (see *Figure 3.7*).

3.1.4 Glow-discharge treatment of carbon films

It has been found that carbon support films are inherently hydrophobic, but some think that this is not the case with freshly prepared carbon films. I have routinely used carbon films prepared by dissolving away colloidon or formvar using chloroform, without encountering sample or stain-spreading problems that are an immediate and readily visible sign of carbon hydrophobicity. The most commonly used procedure for avoiding hydrophobicity is to treat the carbon surface by *glow discharge* under partial vacuum (e.g. 2×10^{-1} Torr) in an appropriate small bench-standing commercial apparatus or in a larger vacuum coating apparatus with a glow-discharge attachment [note that this equipment generally uses a high voltage input (Namork and Johansen, 1982); but see Aebi and Pollard (1987) for an alternative]. It should be borne in mind that apart from removing surface contaminatory hydrocarbons, the gaseous ionic bombardment during the glow discharge is likely to generate both positive and negative charges on the carbon surface. This makes the carbon surface highly reactive and adsorptive, a feature that may or may not be desirable (remember *activated* carbon has powerful adsorption properties for many chemicals). If glow discharge is performed under a partial atmosphere containing an organic amine such as amylamine (Dubochet *et al.*, 1971), the carbon surface will become uniformly positively charged, making it suitable for the firm attachment of negatively charged proteins and nucleic acids. Strangely, it is maintained that glow discharge in the presence of nitrogen tends to give a mixture of positive and negative charge. Advice about the length of time that this treatment should be applied varies. I normally use 15–20 sec, but others give much longer treatments, up to several minutes. Prolonged glow-discharge times may generate an extremely strong adsorptive force which can then interact with any biological material applied to the carbon. This powerful adsorption may selectively orientate molecules or even promote partial flattening prior to the addition of the supportive negative stain and prevent stain penetration between the biological material and the carbon support film. It is general practice to use glow-discharge treated carbon films or carbon–plastic films within 30 min of preparation, as the increase in hydrophilicity/decrease of hydrophobicity is not permanent. It should be noted that the hydrophilic properties of carbon and carbon–plastic films differ somewhat. In general carbon–plastic films retain a considerably deeper layer of negative stain than carbon alone. This has obvious advantages in some instances, but often leads to the production of excessive mass thickness and unacceptable irradiation sensitivity, particularly for the carbohydrate-containing negative stains.

If one deliberately chooses to avoid adsorption to a carbon support, glow discharge should not be used. Freshly prepared grids or chloroform-washed grids may be suitable; treatment of the carbon surface with a low concentration (e.g. 1 mM) of n-octyl-β-D-glucopyranoside in water prior to the addition of the biological specimen is likely to block or at least considerably reduce the adsorption force, thereby allowing greater freedom of rotational mobility of a sample prior to drying of the negative stain.

An alternative to glow discharge, introduced by Barnakov (1994) was to sequentially treat carbon films with 1% sodium phosphotungstate at pH 6.0 in order to improve the adherence of biological samples to hydrophobic carbon, prior to a final negative staining with uranyl acetate. The sequence of treatment appears a little strange in this instance, since the phosphotungstate treatment is applied *after* the initial attachment of the biological material, but an interaction of phosphotungstate with natural and artificial membrane systems and proteins has been encountered.

3.2 Negative stains

3.2.1 An introduction to the properties and uses of negative stains

It is perhaps no coincidence that the first negative stains were derived from positive staining salts already in routine use within the EM laboratory, although as negative stains they were used under significantly different conditions. Ethanolic uranyl acetate and phosphotungstic acid were in general use in the 1950s as positive stains for tissue sections, spread chromatin and nucleic acids. For negative staining, uranyl acetate was used as a low concentration *aqueous* solution (without adjustment of pH since it readily precipitates on neutralization), but phosphotungstic acid was usually neutralized to remove its pronounced positive staining properties. The likelihood that aqueous uranyl acetate still retains the property of positively staining biological material was not addressed initially; however, the fact that the negative staining procedure is usually rapid suggests that significant positive staining is unlikely, but it cannot be dismissed (cf. uranyl acetate stabilization; see Section 3.2.2).

The most widely used negative stains are listed in *Table 3.1*, with some

Table 3.1: Commonly used negative staining salts

Uranyl acetate: 1% or 2% aqueous[a] solution (pH 4.5)
Uranyl formate: 1% aqueous solution (adjust to pH 4.5–5.0 with NaOH)
Sodium/potassium phosphotungstate: 2% aqueous solution (adjust to pH 5.0–8.0 with NaOH)
Sodium silicotungstate: 2% aqueous solution (adjust to pH. 5.0–8.0 with NaOH)
Ammonium molybdate: 2% aqueous solution[a] (adjust to pH 5.3–8.0 with NaOH/NH$_4$OH)

The neutral surfactant n-octyl-β-D-glucopyranoside (OG), M_r 292.4, which possesses the high critical micelle concentration (CMC) of 25 mM, can be added at a concentration of 1 or 2 mM to improve the spreading and penetration properties of the stain.
[a] It is possible to prepare negative staining solutions in low concentrations of ethanol in water, should this be desirable. The surface tension of the stain solution can be reduced by ethanol or by the addition of OG.

Table 3.2: Some lesser used negative staining salts

Uranyl nitrate
Uranyl sulphate
Uranyl acetate–EDTA complex
Uranyl magnesium acetate
Ammonium uranyl-oxalate[a]
Sodium tungstate
Methylamine tungstate (particularly useful for viruses)
Sodium phosphomolybdate
Sodium tetraborate
Cadmium iodide
Cobalt nitrate–ammonium molybdate
Thalium fluoride
Potassium phosphate
Silver nitrate
Vanadium molybdate
Methylamine vanadate (Nanovan)

---

[a] Uranyl-oxalate is a complex anion; pH can be adjusted to neutrality, but unstable and photolabile.
Note, many of the above negative stains have a rather low density, which may be of advantage in certain cases, such as for the localization of small gold particles bound to proteins. Methylamine vanadate is claimed to be particularly useful in this respect.

less frequently used stains given in *Table 3.2*. It is possible to assess the likely negative staining potential of the different salts on the basis of the atomic number Z of the heavy metal atom they contain, which together with the size of the hydrated ion involved will generate the ionic and solution density. Perhaps of greater practical significance is the actual aqueous concentration of the negative stain to be used, together with the final thickness of the fluid layer on the carbon film and therefore the thickness of the dried stain layer, which will determine the net mass–thickness ρt (mass per unit area A of specimen) of the negative stain as it surrounds and embeds any biological structure (note, mass = density × volume, ρV =

ptA). In this case, deficiency contrast created by differential electron scattering (primarily elastic scattering) due to the difference in the atomic composition of the biological material versus the surrounding negative stain (sometimes termed mass–thickness amplitude contrast) is the main imaging principle, combined with some defocus-induced phase contrast. Cryoelectron microscopy of unstained vitrified biological specimens, at an optimal thickness of vitreous ice where the mass–thickness of the biological material is slightly greater than that of the surrounding ice, has indicated that image formation is not primarily dependent upon amplitude contrast (see Chapter 6). Here the small differential elastic electron scattering between ice and biological material, together with the more significant defocus phase contrast (but with increasing fluctuation of contrast transfer function (CTF) with defocus), both play a role in image formation (Misel, 1978, and see Section 7.5). In negative staining, however, amplitude contrast does play the major role in image formation and images can be recorded close to focus, within the plateau of the instrumental CTF, whilst still retaining considerable contrast, even with low stain concentrations.

The cationic uranyl negative stains undoubtedly impart the greatest amplitude contrast, but they tend to have a slightly larger microcrystalinity/granularity than the other stains after drying. Of the uranyl negative stains, uranyl formate appears to produce the finest granularity, but it presents slightly greater difficulties for routine use. The ammonium uranyl–oxalate (anionic) complex usually produces rather coarse image granularity and although slightly unstable/photolabile it has the benefit of being neutralizable. Despite producing somewhat lower image contrast than the uranyl salts, the anionic Na/K-phosphotungstate and -silicotungstate negative stains, both of which can be neutralized, generate a fine granularity and have become widely used negative stains. Although ammonium molybdate was introduced initially because of the beneficial isotonicity of a 2% solution for studies on mitochondria (Munn, 1968; Muscatello and Horne, 1968) its continuing use in recent years is primarily because of its fine granularity after drying, together with potentiation of 2-D array/crystal formation (see Section 3.5.1). Almost all the main heavy metal-containing negative stains contain complex polycations or polyanions, the chemistry of which is known to be highly intricate, usually dependent upon the pH of the solution together with 'ageing' effects and photolability (Tranum-Jensen, 1988). Thus, although it is difficult to establish the detailed physical chemistry of these negative stain solutions, a wealth of practical experience has been accumulated from which some salient points will be expanded below. Interestingly, to the practised eye, it is readily possible to distinguish between the different negative stains in the EM images, since image granularity and spreading properties of the stains differ in a reproducible manner.

The particulate biological samples to which negative staining has been widely applied over the past 40 years tend to fall into a few discrete categories, namely viruses (Horne and Wildy, 1979), macromolecules and macromolecular assemblies, naturally occurring and reconstituted membranes and lipid suspensions, together with bacterial membranes and cell wall structures. Several examples of these applications will be given in Chapter 5 and they serve to indicate the broad range of usefulness of negative staining in biology and medicine. To further emphasize the possibilities for negative staining it might be appropriate to mention that it can even be used to study the structure of aqueous suspensions of polymer particles (Harris *et al.*, 1995a) and no doubt other synthetic particulate materials, providing they are spread as a thin layer and are not totally hydrophobic in character. Even in these cases, however, addition of neutral surfactant can be of assistance to improve stain spreading over the carbon support and around the adsorbed hydrophobic particulates.

3.2.2   The commonly used negative stains

Uranyl acetate. This negative staining salt is widely used in the study of biological samples, and is usually used as a 1% or 2% (w/v) solution in double-distilled water without adjustment of the initial pH of approximate pH 4.3–4.5. Precipitation of uranyl acetate rapidly occurs on the addition of alkali, when pH 5.0 is exceeded. Some prefer to store this stain in dark bottles due to slight photolability, but I have not encountered any problem with solutions stored in clear bottles away from direct sunlight. Room temperature storage for this and other negative stain solutions is usually appropriate, although refrigerator storage can be employed. Uranyl acetate solutions can be filtered or centrifuged immediately prior to use if freshly made, but a more important and widely general factor with stored negative stain solutions is *never* to invert the storage bottle or mix the solution before use. It should be borne in mind that all uranyl salts possess a low level of inherent/natural radioactivity and reasonable precautions should be taken with the stock of solid and with the disposal of solutions and contaminated consumables under the guidance of a local Radiation Protection Officer. When air-dried as a thin layer, uranyl acetate possesses a microcrystallinity/granularity claimed to be in the order of 3 Å (Shlomo Trachtenberg, personal communication) but the presence of buffer salts may also tend to increase the size of uranyl microcrystals and there is evidence for increasing size of the microcrystals and even of stain mobility during electron irradiation (Unwin, 1974). The acidity and positive staining potential of uranyl acetate are features that should be borne in mind when using this stain. These may stabilize the structures under investigation in some instances, but cause severe aggregation or precipitation in others. It should be noted that the use of phosphate buffer solutions is not compatible with uranyl negative staining, since an insoluble precipitate of uranyl phosphate forms. I have found it beneficial to include 1 mM n-octyl-ß-D-glucopyranoside in uranyl acetate negative stain solutions. This reduces precipitation in the stock stain solution during storage and improves spreading and penetration within the biological material (Harris and Horne, 1994; Kühlbrandt, 1992). The pH of negative stains is not constant during drying; for the uranyl salts there is a slight but significant decrease in pH as saturation is reached (Tranum-Jensen, 1988).

Uranyl formate. Generally, this negative stain is used at a concentration of 1% (w/v) or slightly lower. As uranyl formate possesses the highest available density (Bremner *et al.*, 1992) and generates extremely fine granularity following drying, it is considered to be particularly useful for the study of small protein molecules. It is recommended that uranyl formate should be dissolved in carbon dioxide-free double-distilled water (produced by boiling). It dissolves rather slowly and should be stirred in the dark and subsequently membrane filtered (0.22 µm). The solution initially has a pH of approximately 4.0 and it should be adjusted to pH 4.5–5.0 with concentrated NaOH or NH_4OH. As with uranyl acetate, further pH adjustment towards neutrality leads to precipitation of uranyl formate. Due to photolability, uranyl formate should be stored in brown bottles or clear bottles wrapped in aluminium foil. In solution, this negative stain is not stable over extended periods, although it can be stored for lengthy periods when frozen.

Sodium phosphotungstate and sodium silicotungstate. For convenience, these two negative stains, which possess somewhat similar properties, will be dealt with together. Both materials are available as free acids and need to be titrated with NaOH (or KOH) to the desired pH. Usually, 2% (w/v) solutions should be used. These are stable at room temperature during long-term storage. Bacterial growth has, however, been detected in pH 7.0 phosphotungstate solutions, possibly encouraged by the release of phosphate due to slow degradation. Bacterial contamination can be prevented by the addition of a small volume (e.g. 100 µl) of chloroform to the stain solution. The available evidence indicates that dried sodium silicotungstate possesses a somewhat finer microcrystallinity than sodium phosphotungstate, but both salts are suitable as negative stains for a wide range of biological material. A major note of caution should, however, be observed. There is much evidence that phosphotungstate will interact with lipoproteins, biological membranes and lipids to generate artefacts and/or structural changes. Considerable care should therefore be observed under these circumstances, and for comparison purposes a range of different negative stains should be employed.

Ammonium molybdate. Ammonium molybdate was introduced as a negative stain primarily because of the benefits it provides for the study of membrane suspensions in isotonic solution. It has, however, continued to be used because of its broad acceptability for the negative staining of viruses, liposomes and macromolecules, as well as cellular membranes and bacterial surface layers. Although ammonium molybdate has a rather low density and therefore generally creates a low mass–thickness on drying, it has a fine microcrystallinity/granularity, but usually only in the complete absence of any traces of buffer salts and NaCl. Routinely, 2% (w/v) solutions of ammonium molybdate are employed. On dissolving the salt, ammonium molybdate has a pH of approximately pH 5.3 and it can be titrated with NaOH or NH_4OH to the desired pH, usually in the range of pH 6.0–7.0. Some instability of pH will be encountered and the pH of solutions should be readjusted after a few days. Significantly, structural differences in proteins have been detected with ammonium molybdate, dependent upon whether NaOH or NH_4OH is used to neutralize the stain (Peters *et al.*, 1992). Whilst for most biological materials ammonium molybdate appears to have no direct interaction as a positive stain, there is an increasing list of materials that are labile in ammonium molybdate solutions, particularly at lower or higher pH values rather than at neutrality. Molybdate certainly has the capacity to sequester protein-bound divalent cations which could cause virus and protein dissociation (Harris *et al.*, 1995b; Horne, 1986) but it may also act directly as a mild chaotropic agent, thereby producing controllable and experimentally useful biochemical changes to biological structures.

3.2.3 The less frequently used negative stains

Of the less frequently used negative stains listed in *Table 3.2,* only brief comment will be made, since there are few examples of these stains providing any additional benefits. *Methylamine tungstate*, as a 2% (w/v) solution has been shown to be useful for the negative staining of viruses and membranes, and has also been used for macromolecules (Oliver, 1973). This negative stain tends to spread evenly and remains as a somewhat deeper fluid film, and therefore thicker dried layer, on the carbon support film than most other stains, thereby providing an increased mass–thickness of dried stain embedding the biological material. The full benefits of being able to control the thickness and therefore the embedding properties of the negative stain layer have not been fully addressed until recently (Harris and Horne, 1994). Thus, it is possible that methylamine tungstate will find an increased usage in future years, either alone or in combination with the carbohydrate trehalose.

A recently developed lower density negative stain *methylamine vanadate* (Nanovan), is claimed to produce a much smoother background granularity than sodium phosphotungstate and to be very stable in the electron beam. The lower density and electron scattering power of methylamine vanadate also permits a thicker supportive layer of stain to be produced (Hainfeld *et al.*, 1994).

A neutral pH solution of *ammonium uranyl oxalate* can be produced by titrating a mixture of uranyl acetate and oxalic acid with NH_4OH until a pH of 6.5–6.8 is reached (Mellema *et al.*, 1967). Thus the presence of the high density uranyl-oxalate anion combined with the neutrality of the so-lution can provide a benefit in some instances. Unfortunately, the granu-larity of the dried stain is rather excessive and the solution is extremely photolabile. Storage in the dark or in the frozen state is recommended.

Sodium tetraborate is a rather unique low-density negative staining salt. When used as a 2% solution, sodium tetraborate enables a clear dis-tinction to be made between ferritin and apoferritin. This is because the dried negative stain has a lower mass–thickness than the iron hydroxide core of ferritin, but has a greater mass–thickness than the protein shell of this molecule (Massover, 1975, and Chapter 5). It is, however, likely that a similar density could be produced from lower than usual concentrations of the other negative stains, but this may produce a very thin dried layer of stain.

The combination of *cobalt nitrate* and *ammonium molybdate* provides an interesting negative stain for the study of liposomal suspensions, since the cobalt cation and molybdate anion participate in the colloidal 'tri-com-plex' reaction, to produce bilayer interactions and phospholipid flocculation.

3.2.4 Carbohydrates and carbohydrate-negative stain mixtures

The density difference between biological material containing a mixture of protein, lipids, nucleic acids and carbohydrates, and pure carbohydrates such as glucose or trehalose is rather small. Nevertheless, it has proved possible to image and to obtain electron diffraction patterns from thin 3-D protein crystals and 2-D crystals of membrane proteins under conditions of low temperature and strict low electron dose. Crystallographic image processing (Section 10.2) has then enabled high resolution (3.5–7.0 Å) in-formation to be recovered in several instances. The large number of hydrophilic -OH groups appear to protect protein molecules during drying and embedding in carbohydrates. Because of the limiting contrast avail-able, the imaging of single molecules has not been pursued greatly using pure carbohydrates, although some limited success has been achieved with aurothioglucose and cadmium thioglycerol (*Table 3.3*). The former has been used to study the structure of ribosomes and RNA location, and the latter to study annelid haemoglobin molecules as well as bacterial surface lay-ers.

Mixtures of uranyl acetate and glucose have been used for the negative staining of protein crystals by Kiselev *et al.* (1990), who from electron dif-fraction claimed a resolution of 5 Å. Under conventional illumination con-ditions, glucose is extremely unstable in the electron beam, readily pro-ducing specimen 'bubbling' even in relatively thin negative stain–glucose mixtures. Trehalose has, however, been found to possess considerably greater stability (Harris and Horne, 1994; Harris *et al.*, 1995c), at the level of 1% (w/v) in 5 % (w/v) ammonium molybdate or 4% (w/v) uranyl acetate. The inclusion of trehalose necessitates an increase in the concentration of

Table 3.3: Carbohydrates and carbohydrate-containing negative stain solutions

---

Glucose
Trehalose
Aurothioglucose
Cadmium thioglycerol
Negative stain + glucose
Negative stain + trehalose[a]
Tannic acid (tannin) (M_r 1701)[b]

---

[a] 4% or 5% (w/v) negative stain + 1% trehalose provides satisfactory contrast; lower stain levels may also be satisfactory for high resolution studies and for the location of small gold probes. The overall stain density can be readily *tailored* by adjusting the negative stain %, whilst maintaining 1% trehalose.
[b] Has a stabilizing interaction with proteins.

the negative staining salt beyond the more usual value of 2% (w/v) because of the 'diluting' effect of the carbohydrate on the overall mass–thickness of the dried stain. Excessively thick layers of negative stain containing trehalose are subject to rapid electron beam-induced 'bubbling', and should therefore be avoided (see Chapter 4). This also applies to samples cooled to −175°C, or lower, in a liquid nitrogen-cooled specimen holder, and although the thinner regions of stain are more clearly protected at this temperature, strict low electron dose studies are desirable. In general, it has been found that ammonium molybdate–trehalose is somewhat more stable in the electron beam than the uranyl acetate–trehalose combination.

Another carbohydrate of interest to electron microscopists is tannic acid (M_r 1701). Indeed, tannic acid has been used for many years as a stabilization agent for cytoskeletal proteins during tissue processing. Tannic acid has useful biochemical properties for enzyme and protein immobilization and adsorption, and forms insoluble complexes with proteins, properties that correlate with its beneficial fixative action on tissues. Recently it has found application as an embedding agent for 2-D and 3-D crystals of soluble and membrane proteins and tubulin (Akey and Edelstein, 1983; Nogales *et al.*, 1995; Wang and Kühlbrandt, 1991) from which resolutions in the order of 6.5 Å have been achieved. As with glucose, it would appear that tannic acid will continue to be used mainly for the study of crystalline material, rather than single particles. Nevertheless, here too the possibility exists for creating tannic acid–negative stain mixtures which may provide a useful combination for stabilizing and embedding dispersed particulate biological samples, as well as protein crystals.

The stabilization of biological material prior to negative staining by fixation with glutaraldehyde has been routinely used by some investigators (Tranum-Jensen, 1988) although most prefer to employ negative staining *without* the addition of fixative to biological samples, unless this proves to be absolutely necessary. If used, fixation can be performed in solution before spreading the biological sample on the carbon support film, with a low concentration of glutaraldehyde (0.01–0.1% w/v). It should be remembered that glutaraldehyde fixation is not compatible for materials suspended in Tris-HCl buffer solutions, which should be removed by dialysis before the addition of glutaraldehyde. The reaction with glutaraldehyde can be blocked by the addition of glycine. After fixation, both buffer salts and glutaraldehyde can be removed by dialysis against distilled water. Alternatively, glutaraldehyde can be applied as a droplet of low concentration solution (0.1–1.0% v/v) to biological particulates *already* adsorbed on to a carbon support film, with incubation for a short period of time, followed by droplet water-washing (see Section 3.3.1).

3.2.5 Spreading agents

The need for the inclusion of a spreading agent during negative staining was initially addressed by virologists, who encountered severe problems with purified virus samples, whereas impure samples containing a range of soluble and membranous components spread easily. Gregory and Pirie (1973) included the antibiotic bacitracin (M_r 1411), which assisted the situation and is still used for this purpose (see also Section 3.3.1). Bacitracin is not the best additive, because of its rather high molecular weight and the neutral surfactant n-octyl-β-D-glucopyranoside (octylglucoside/OG; M_r 292.4) is considered to be a better alternative. This surfactant, which has a high critical micelle concentration (CMC) (25 mM), can be added to the biological sample to a concentration of a few mM, or to the negative stain solution (see below and Section 3.2.2). Other agents such as octadecanol (M_r 270.5) and the low molecular weight polyethylene glycols are also considered to possess useful possibilities (Harris and Horne, 1994), see *Table 3.4*.

Despite the above, it is my experience (from studies on subcellular fractions and purified protein molecules) that it is rarely necessary to include a spreading agent in the biological sample; however, it may prove useful to routinely include OG in negative stain solutions to improve stain penetration and to induce a more evenly spread film of fluid on the carbon

Table 3.4: Some spreading and 2-D crystallization agents used in negative staining and associated techniques

Bacitracin
Bovine serum albumin (BSA)
Ethanol
Octadecanol
n-Octyl-β-D-glucopyranoside (OG)
Polyethylene glycol (PEG)[a]

[a] 2-D and 3-D protein crystallization agent.

film. Usually, glow-discharge treatment of support films eliminates the need to include a spreading agent. Should glow-discharge treatment be unacceptable, for instance because of the undesirable selective orientation of a protein on the carbon film that adsorption has produced, then inclusion of a low concentration of OG within the sample may be contemplated or alternatively droplet washing the carbon surface with a 1 mM OG solution, *prior* to application of the sample, may be helpful to block or reduce the adsorption forces.

3.3 The droplet technique

The droplet procedure to be presented below represents the simplest negative staining technique (Harris and Agutter, 1970; Harris and Horne, 1991). It can be understood and performed in a matter of minutes by students and researchers alike. The alternative, related approach, of directly pipetting 5 or 10 µl volumes of specimen solution, water and negative stain solution consecutively on to carbon support films, whilst still used by some, is not considered to be as simple. Although it is possible to mix sample and negative stain prior to application to the carbon support film, this is not the usual approach due to the possibility of undesirable stain–material and buffer salt–stain interactions (see Sections 3.3.2 and 3.5.5). However, some maintain that this approach is useful (Prakash Dube, personal communication). It may reduce the binding of protein to the carbon film and thereby generate a more homogeneous/isotropic range of protein images that are not flattened or partly stained due to stain exclusion at the point of adsorption. If performed, one should ensure that the buffer components are compatible with the negative stain; it is likely that the best results would be derived from a sample solution in water or very dilute buffer solution. With samples that have been water dialysed or greatly diluted (e.g. 1:100) with negative stain from a high protein concentration (thereby diluting the buffer salts), sample–stain mixing may prove to be routinely satisfactory. For the production of rapidly frozen samples from thin aqueous films of sample in the presence of negative stain across holey carbon support films (i.e. without a requirement for adsorption to a carbon layer), sample–stain mixing may be the only way to proceed, but see Section 3.3.1 and Cyrklaff *et al.* (1994).

In view of the currently increasing awareness of viral infectivity, protein immunogenicity and the general toxicity of reagents in the EM laboratory, especially as aerosols, the application of samples, sample material mixed with negative stain, and negative stains by spraying (the spray-droplet approach) will not be presented.

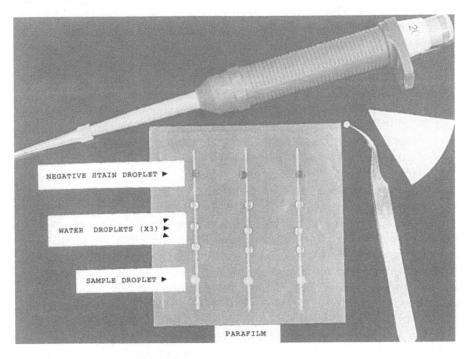

Figure 3.8: Droplet negative staining.
A typical layout of materials and equipment required for the single droplet negative staining procedure.

3.3.1 The standard droplet technique

This technique, used principally by biochemists and cell biologists since the early days of negative staining, will be presented as a slightly modified and more detailed version of that described by Harris and Horne (1991). *Figure 3.8* shows a simplified layout of the droplet negative staining technique, for the preparation of three specimens in this instance. In my experience it is most convenient to restrict the number of specimens produced in any one batch to a maximum of 10, to avoid possible confusion and inadvertent touching of the droplets.

Procedure. An appropriate length of Parafilm should be cut and loosely attached to the bench at its sides and ends by drawing a blunt object in a straight line across the paper overlay. Several more evenly spaced lines should then be produced across the strip, depending upon the number of specimens to be prepared, and the paper overlay removed leaving the Parafilm attached to the bench. Twenty microliter droplets of water, negative stain and sample solution are then successively positioned on the individual Parafilm lines. The number of water droplets can be anything from one to five, depending upon the salt or solute concentration within the specimen solution that needs to be removed before applying the negative stain droplet. The protein concentration in the sample should be in the order of 0.05–0.5 mg ml^{-1}, depending upon the molecular mass of the macromolecule or particle under investigation. In general, the higher the mass the higher the maximum possible concentration before molecular overloading by superimposition occurs. For membrane suspensions a protein concentration of 0.2–0.5 mg ml^{-1} is usually satisfactory and for lipid and liposomes suspension 0.5–1.0 mg ml^{-1}. Fine tipped curved forceps are most suitable for handling grids during negative staining. Some very skilful operators prefer to bend the tips slightly together, to reduce the possibility of sample fluid being drawn up between the two inner sides by capillary 'wicking' action. This is certainly a hazard one needs to be aware of, since this trapped fluid can flow back on to the grid and mix with the negative stain at the final drying stage.

Using a specimen grid coated with an appropriately glow-discharge treated carbon or carbon–plastic support film, held precisely at the continuous ring of metal at the edge (an easily sliding ring of silicone rubber tubing on the forceps handle is useful; some prefer to use reverse action forceps) the grid should be touched to the sample droplet and removed. After allowing a period of time for adsorption of the sample to the carbon, which can be anything between 10 sec to 10 min, depending upon the available protein concentration (longer times for very low protein concentrations) the fluid should then be removed by carefully touching on to the edge of a filter paper wedge. The filter paper can then be touched to the tip of the forceps to check whether any 'wicking' of fluid has occurred. For very dilute specimens it is possible to concentrate particulate material directly on to a carbon film by placing a grid at the base of a modified microcentrifuge or airfuge tube and centrifuging the sample on to the carbon (Schwartz *et al.*, 1983), but this is not usually necessary.

Before the sample fluid has had time to dry, the adsorbed material should be rapidly washed sequentially with one to five (or more) droplets of distilled water, removing the fluid in the same manner as for the sample. A sample buffer, salt and other solute of concentration up to 500 mM or more can readily be removed by five droplet washes, providing the fluid is removed as efficiently as possible at each wash. Finally, the droplet of negative stain solution is touched and likewise removed, and the grid allowed to air-dry. With the negative stains containing glucose or trehalose it is important to maintain contact between the filter paper and the carbon surface for a sufficient time to remove maximal fluid, otherwise an excessively thick layer of dried stain may be produced.

An alternative approach to the droplet technique is the use of a bacteriological platinum loop for grid washing and negative staining by floating the grid sample-side-down on 100 µl droplets on Parafilm or water in the wells of a ceramic or polypropylene spotting plate (*Figure 3.9*). The grid is finally positioned on to a filter paper from the inverted bacteriological loop so that the sample is now uppermost. This will give a very

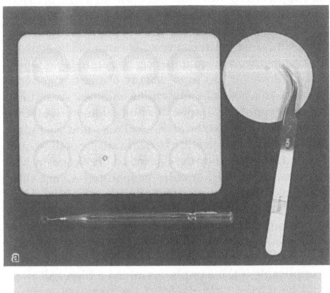

Figure 3.9: The bacterial loop procedure for washing and negative staining of EM grids.
This procedure can be carried out in either the wells of a spotting plate (a) or 50–100 µl droplets on a Parafilm strip. Grids are floated successively on the water and negative stain, finally placed 'carbon-up' on a filter paper, and then removed with forceps for drying. The bacteriological loops (approx. 4 mm in diameter) can be prepared from 0.2 mm diameter platinum wire inserted and sealed into a drawn-out molten glass tip of a pasteur pipette. The handling of a specimen grid on such loops will be easily performed. (b) Shows platinum loops with and without a grid suspended on a droplet of fluid. (Procedure courtesy of Milan Nermut.)

gentle radial removal of the excess stain, without imposing any directionality of spreading (Milan Nermut, personal communication). With fine forceps the grid is removed from the filter paper and after air-drying, the specimen grid is ready for immediate study in a TEM.

A hazard of droplet negative staining that is sometimes encountered is the spreading of sample, washing fluid and stain on to *both* sides of the carbon, even though only one side of the carbon initially touches the sample droplet. This becomes obvious as the water washings and stain can be seen on both surfaces of the grid as a *mini-biconvex lens*. Usually such grids will give somewhat anomalous results, and should be discarded if sufficient sample is available to repeat the specimen preparation. This problem may be caused by a partially broken carbon film, or by holding the grid rather too far into the tips of the forceps (i.e. too far over the continuous metal edge) (see Chapter 4).

Droplet negative staining on holey carbon film. In an attempt to produce specimen material supported by and embedded within a film of dried negative stain alone, several investigators have tried to spread the biological sample and negative stain across small mesh bare grids and holey carbon films. This approach has had limited success because of the considerable instability of the thin *unsupported* sample–negative stain layer when in the electron beam. It is, however, an area worthy of further serious investigation since it avoids any possible adsorption-induced structural change to the biological material (see Chapter 9). Fujiyama *et al.* (1990) used this approach for the study of bacterial F_1–ATPase, and included bacitracin (20–40 µg ml⁻¹) in 2% uranyl acetate to facilitate spreading and render the stain film over the holes more resistant to breakage in the electron beam. Nevertheless, these workers also evaporated a very thin layer of carbon on both sides of the specimen grid *after* the negative stain had dried. It is very likely that the inclusion of trehalose in the negative stain would also improve the spreading and stability of the dried film across the holes, but care would need to be exerted to maintain minimum electron dose to ensure that the stain film remains stable. Obviously, this approach might not be applicable if the sample contained a high salt concentration, but it should be particularly useful if the negative stain + trehalose can be mixed directly with the biological material (in low salt buffer or distilled water) before applying to the holey carbon film. The fact that viruses and protein molecules remain held at the fluid–air interface during washing (Cyrklaff *et al.*, 1994; Johnson and Gregory, 1993) indicates that on-grid water washing on a holey carbon film might nevertheless be used, prior to negative staining.

Grid storage. Negatively stained specimen grids can most conveniently be stored in plastic grid-boxes with a sliding or rotating lid (*Figure 3.10*). Be aware of static problems, which may prevent insertion of grids or cause them to spontaneously eject when the box is opened; static charge can usually be removed by carefully running a finger across the surface of the open sliding-lid box. Ensure that the sliding lid can easily be moved in both directions; limited lubrication of the edges of the box with 'finger grease' or a trace of silicone fluid may assist matters with a difficult lid. An alternative type of storage box with a clip-on hinged lid and a moulded rubber insert containing several small compartments, is also available commercially; it is convenient for horizontal laboratory storage, but unlike the sliding-lid type it is not suitable for pocket or briefcase transport, or postage of grids to other laboratories and these boxes are less satisfactory for increasingly large numbers of stored specimens. Specimens negatively stained with uranyl acetate will be stable and usable for many months, but those stained with ammonium molybdate may show some progressive recrystallization of the initially amorphous/vitreous layer of dried stain. This appears to be particularly bad when traces of buffer salts have remained on the carbon surface and mixed with the negative stain.

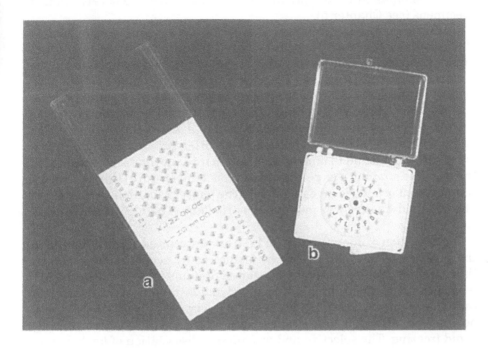

Figure 3.10: Grid storage boxes.
  Examples of convenient grid storage boxes for negatively stained specimens, (a) with a sliding lid and (b) with a rotating lid.

With uranyl acetate and ammonium molybdate specimens containing trehalose, conversion of the vitreous/dried film of stain + carbohydrate into a crystalline state occurs after a period of several weeks. Some prefer to store grid boxes within a vacuum desiccator, which may prevent such recrystallization, possibly promoted by the continually varying atmospheric temperature and humidity.

3.3.2 Dynamic cellular and biochemical experiments

Negative staining can readily be performed on cellular and biochemical samples taken during experimental treatments performed over an increasing time period (e.g. 1 min to 60 min, or several hours to a day) providing the time required to perform on-grid adsorption, droplet washing and the final addition of the negative stain droplet is not itself excessive in relation to the speed of any possible structural change that might be under investigation. Thus, the slow progressive formation of the filamentous complex of the *Escherichia coli* chaperone GroEL (cpn60) with the smaller co-chaperone GroES (cpn10) under conditions of limiting GroES availability has been performed successfully (Harris *et al.*, 1995d). Likewise, the early and later stages of toxin–membrane or toxin–lipid interactions could be monitored by the production of consecutive negatively stained specimens (see Chapter 5).

Although most negative staining is performed at room temperature, it is possible to apply both biological sample and negative stain to support films at lower or higher temperatures, should this be appropriate for the material under investigation. Thus, Benedetti and Emmelot (1968) used negative staining at 37°C in their early study of rat liver gap junctions, and the lipid phase change and vesicularization of sphingomyelin was readily shown by specimen preparation over a wide temperature range (4–80°C) (Harris, 1986a). In addition, lysolecithin crystals which were stable at 4°C, were found to undergo disruption into micellar complexes when the negative staining was performed at room temperature (Harris, 1986b). Similarly, calorimetric investigation of the phase transitions in sonicated phospholipid vesicles was monitored by Kodama et al. (1993) using negative staining.

Reformation of microtubules (MTs) from dissociated α and β tubulin in the presence of GTP can readily be monitored by negative staining, and the attachment of microtubule-associated proteins (MAPs) along the MT length can be defined. In addition, the dynamic temperature-dependent recycling of microtubules (spontaneous dissociation/reformation) can also be assessed, although such studies can be more accurately performed by rapid freezing. The selective and progressive dissociation of keyhole limpet haemocyanin type 2 (KLH2) when incubated in the presence of slightly acidic ammonium molybdate (e.g. pH 6.5–5.5) has enabled the more stable KLH1 to be quantitatively isolated for biochemical characterization (Harris et al., 1995b); see also Sections 3.5 and 5.4. In addition, the slow reassociation of pH 5.7 ammonium molybdate-dissociated KLH2 and high pH (glycine-NaOH, pH 9.6) dissociated KLH1 and KLH2 produced by dialysis against 100 mM Ca^{2+} and Mg^{2+}-supplemented Tris-saline haemocyanin stabilizing buffer has also been monitored over a period of days and even weeks by conventional negative staining using the droplet technique. The direct effects of negative stains on viral particles can readily be monitored by mixing prior to adsorption of virus to carbon support films (Horne, 1986; Nermut, 1982b, 1991).

One further approach that can be applied to lipoproteins and biological membranes adsorbed to a carbon support film is the partial or complete removal of the lipid by organic solvent extraction before negative staining (for details see Phillips and Schumaker, 1989). It may be desirable to stabilize the biological material by fixation in solution with a low concentration of glutaraldehyde (e.g. 0.1% v/v) before application to a carbon support film or after adsorption to the carbon support film. Also, note that the organic solvent extraction is likely to alter the hydrophilicity of the carbon film, which may then severely interfere with the spreading of the negative stain. Inclusion of a low concentration of OG in the negative stain solution may assist, in this instance.

3.4 The floating methods

3.4.1 Carbon film adsorption

Although not currently widely used, this method of negative staining is of considerable technical interest, since it emphasises the natural adsorptive properties of carbon, and provides a logical if not strictly historical link between the droplet procedure (Section 3.3) and the mica–carbon transfer procedures (Section 3.5). The procedure was introduced by the late Robin Valentine, who used it extensively in his studies on enzymes and enzyme complexes (Valentine *et al.*, 1968). There is an immediate similarity with the droplet procedure, in that samples, water and negative stain droplets placed upon a Parafilm surface are employed, as shown in *Figure 3.8*. Alternatively, droplets can be placed in micro-wells in a Teflon block (see Section 3.4.2).

Procedure. For carbon film adsorption it is convenient to use approximately 50 µl droplet volumes of sample, water and negative stain, rather than 10 or 20 µl droplets. Instead of preparing carbon-coated specimen grids in advance, small pieces (approx. 3 mm × 3 mm) are cut from a larger piece of carbon-coated mica (see Section 3.1.1) and whilst holding at the edge with forceps they should be inserted into the sample droplet, carbon film uppermost. The carbon layer tries to float off, but this is prevented by holding the mica back slightly. After a period of time (10–60 sec, depending on the sample concentration) during which sample adsorption to the floating carbon occurs, the mica piece plus carbon is withdrawn from the droplet and excess fluid removed by touching to the edge of a piece of filter paper. By a successive series of insertions into and withdrawal from the water droplets, unabsorbed sample will be washed away. [A fixation step using a low concentration of glutaraldehyde, e.g. 0.1% (v/v), can be included at this stage, if needed, and followed by further water washes.] The carbon film should then be completely floated off the small piece of mica on to the surface of a negative stain droplet. The floating carbon film plus sample and negative stain can then be recovered by carefully positioning a bare grid, dull surface down, directly on to the carbon, and rapidly removing it. Alternatively, the bare grid can be inserted directly into the stain droplet beneath the floating carbon (prior dampening with a small volume, e.g. 5 µl, of negative stain is necessary) and the grid brought vertically upwards from beneath the carbon (see Section 3.5.1). In both cases, excess negative stain should finally be removed by touching to the edge of a filter paper.

   One modification of this procedure is to actually use grids that already possess a thin continuous or holey film of carbon. This will produce a 'sandwich' staining effect (Lake, 1981) because both the sample and stain are trapped between the two carbon layers. Whilst producing an evenly spread layer of negative stain, this modification has the unfortunate and extremely undesirable effect of causing considerable sample flattening because of the forces that come into play during stain drying (Frank, 1989). Similar reservations must be expressed for the so-called 'pleated/folded carbon sheet' technique.

3.4.2 Lipid monolayer adsorption

This procedure for the preparation of negatively stained and unstained vitrified specimens owes its origin to the long-established Langmuir-Blodgett procedure for the production of lipid monolayers on an aqueous surface. Protein molecules injected into the aqueous phase beneath a packed amphiphilic monolayer may be able to interact with charged groups or other ligands on the hydrophilic face of the monolayer. Thus, it is possible to recover the lipid monolayer with bound protein molecules by transfer on to an electron microscope grid covered by a thin continuous film of carbon or a holey carbon film, for negative staining or direct vitrification without any stain. This technique is particularly important because it has proven potential for the production of 2-D crystals of lipid-bound protein molecules (see review by Brisson *et al.*, 1994).

Procedure. The technique can be performed directly using conventionally produced Langmuir-Blodgett monolayer films or lipid monolayers on the surface of droplets in 4 or 6 mm diameter micro-wells, approximately 1 mm deep, in the surface of a Teflon block (*Figure 3.11*) (Jap *et al.*, 1992; Schultz *et al.*, 1990). The protein solution in an appropriate buffer (20 μl) should be placed in each micro-well, and a lipid monolayer created by injecting 1 μl of phospholipid solution (0.5 mg ml^{-1}, in chloroform-hexane, 1:1 v/v) on to the surface of the protein solution. The exact and always variable conditions for 2-D crystal production on the lipid monolayer (i.e. time, temperature, pH, salt and surfactant content, etc.) will not be dealt with here, since they have to be established for each individual protein molecule under investigation. The transfer of the lipid monolayer + attached protein to the EM grid is a standard requirement in all cases. This transfer often presents some difficulties, since 2-D crystal distortion or breakage can readily occur (Kubalek *et al.*, 1991). The direct touching of a continuous carbon or holey carbon grid to the lipid monolayer from above has been reasonably reliable, but a recent improvement, using a microbiological platinum loop to pick up the floating monolayer from above for transfer to grids, appears to be more satisfactory (Asturias and Kornberg, 1995) since distortion of the monolayer is greatly reduced. This technical approach has long been established for the transfer of grids from one water droplet to another for washing and also for the application of a fluid film of photographic emulsion to thin sections on grids, for autoradiography; see also Anderson (1966) for the transfer of a microbiological loop sample to an EM grid, and Sections 3.1.1 and 3.5.1. Insertion of a fluid-moistened

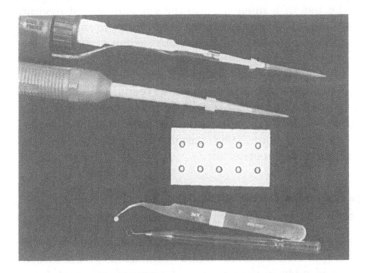

Figure 3.11: Equipment for the preparation of lipid monolayers for 2-D crystallization of proteins.
A Teflon block with 10 micro-wells (wells encircled for photographic demonstration only) and associated equipment for the preparation of lipid monolayers for 2-D crystallization of proteins. The micro-wells are usually approximately 0.5 mm deep and 4 mm diameter. Slightly larger wells can also be used (Asturias and Kornberg, 1995) as this facilitates the transfer of the floating lipid layer and bound proteins, with a bacteriological loop, directly on to bare fine mesh EM grids or holey carbon support films.

holey carbon grid beneath the floating monolayers is also possible and can be followed by further water washing if required, then droplet negative staining, or blotting and rapid freezing (Chapter 7).

3.5 The mica–carbon transfer techniques

As indicated earlier, the origins of the different negative staining, and other preparative techniques such as metal shadowing, and their functional relationship one to another is not always well defined. Thus, the use of cleaved mica to provide a clean hydrophilic surface upon which to spread biological material from a droplet or by spraying was well established in the 1950s. Such preparation has often been performed in the presence of glycerol and volatile buffers such as ammonium acetate, and in combination with freeze-drying. The deposition *in vacuo* of a platinum–carbon or carbon layer on to the dried material can be followed by removal of this replica layer, as during the preparation of carbon support films (Section 3.1). Metal shadowed biological material is in fact adsorbed to the metal–carbon replica and for some cellular studies (e.g. freeze–fracture) it

can be removed by hypochlorite, sulphuric acid or hydrofluoric acid digestion. Nevertheless, the use of a mica surface for spreading and drying
particulate biological material in the presence of ammonium molybdate,
coating with carbon, followed by the flotation of both carbon plus adsorbed
material on to a negative stain solution, was introduced as a standard
procedure in 1974 by Bob Horne and Ivonne Pasquali-Ronchetti, and
termed the 'negative staining–carbon film' (NS–CF) technique. Details of
this technique and a number of variants will be given below, but it is
worth mentioning that others have independently developed somewhat
similar approaches (e.g. Spiess *et al.*, 1987) possibly being unaware of the
earlier NS–CF technique and its potential.

3.5.1   The negative staining–carbon film (NS–CF) technique

During their research on plant viruses, Horne and Pasquali-Ronchetti
(1974) found that when purified icosahedral and filamentous viruses were
spread on freshly cleaved mica *in the presence of* ammonium molybdate,
the viruses would very often adopt ordered 2-D arrays during drying. This
mono-viral layer arrangement was preserved by adsorption to a directly
deposited thin carbon film, which in turn was released on to the surface of
a second negative stain solution. Remarkably, the rehydration and second
drying, usually in the presence of uranyl acetate, appeared to have no
undesirable effects on viral structure. It should, however, be stated that
this aspect of the technique has repeatedly concerned investigators wishing to be sure that biological structures were not altered or even destroyed
during this specimen preparation procedure. Concern has also been expressed that the 2-D arrays may not contain genuine 2-D crystal order,
and simply represent a close-packed monolayer of viruses (Horne, 1977;
Steven *et al.*, 1978).

The adoption of this procedure for areas other than plant virology was
rather slow and although I attempted the procedure from the mid-1970s
for investigations on protein molecules (Harris, 1982; Harris and Kerr,
1976), limited success was achieved. A significant advance came when
Wells *et al.* (1981) included polyethylene glycol (PEG) along with the ammonium molybdate. This greatly improved the 2-D crystallinity of viral
specimens and was also found to assist the 2-D crystallization of a number
of protein molecules spread on mica during the NS–CF procedure (Ghiretti-
Magaldi *et al.*, 1985; Harris, 1991a; Harris and Holzenburg, 1989). Thus,
a system became available within which the pH, concentration of protein/
virus, ammonium molybdate and PEG, and the drying conditions on mica
could easily be varied in order to obtain the best conditions for 2-D crystallization. It should also be emphasized that the NS–CF technique yields
very satisfactory specimens containing randomly dispersed single viral
particles and macromolecules, equivalent and often superior to those produced by the droplet and floating techniques. Strangely, little further ap-

plication of this procedure for the production of 2-D crystals of viruses has occurred since the early 1980s, the field of viral structure being dominated by single particle studies, increasingly from unstained images in vitreous ice (see Chapters 5, 8 and 10).

The first method to be presented below is essentially the same as that previously given by Harris and Horne (1991) with minor modifications. A Teflon block micro-well version of the NS–CF technique will then be described, which is considerably more economical in the use of negative stain solution. The final negative stain solution used in the NS–CF technique can be any of those described above (Chapter 3.2). I most often use either 2% (w/v) uranyl acetate containing 1 mM OG, or 4% (w/v) uranyl acetate containing 1% (w/v) trehalose.

Procedure. The equipment and materials required for the NS–CF technique are shown in *Figure 3.12*. Small pieces of freshly cleaved mica, approximately 0.5 cm × 2.0 cm (pointed at one end) can conveniently be used, but there is no restriction on the use of somewhat larger or smaller mica pieces. The size shown will generate approximately eight to 10 specimen grids. The purified protein solution or viral suspension at a concentration of approximately 0.5 mg ml^{-1} should be selected initially. A lower concentration is likely to be necessary for small molecular weight proteins and for viruses and very high molecular weight proteins, a concentration in excess of 1 mg ml^{-1} may be needed. The protein or virus should be mixed

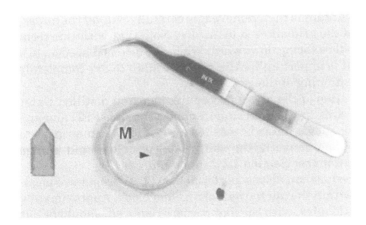

Figure 3.12: A typical layout of the materials used for the final stage of the negative staining–carbon film (NS–CF) technique.

The floating film of carbon + adsorbed virus or protein (arrowed) is released on to the negative stain or water surface within a small petri dish, and individual bare grids brought from beneath, as for the production of single carbon support films (*Figure 3.2*). The small piece of mica (M) remains at the bottom of the petri dish. Note that with this procedure the negatively stained sample material is adsorbed on to the carbon surface facing towards the metal of the specimen grid (*opposite* to that of the droplet negative staining procedure).

with an equal volume of 2% w/v ammonium molybdate (AM) containing 0.1% (w/v) PEG M_r 1500. The pH of the ammonium molybdate can be adjusted to any reasonable value (i.e. pH 5.5–8.5); in the first instance it is advisable to select a pH close to neutrality. The sample–AM–PEG mixture (10 µl) is then applied to the mica surface and spread evenly with the pipette tip horizontal. Excess fluid is removed from the pointed end of the mica by holding it vertically against a filter paper for approximately 2 sec. The remaining fluid, still spread evenly across the mica, is allowed to dry with the mica horizontal. Room temperature drying in a partially covered petri dish will take approximately 10 min. Part of the mica surface will dry rather rapidly, with gradual production of a deeper pool of more slowly drying fluid, usually at the centre of the mica. The drying time can be extended by placing a damp filter paper alongside the mica, and the temperature of drying controlled by placing the petri dish in an incubator, or in a cool room or refrigerator if low temperatures and greatly extended drying/crystallization times are required.

Following drying of the sample–AM–PEG solution, the mica pieces should be carbon-coated *in vacuo*, as described in Section 3.1.1. The thickness of the carbon should be kept to a minimum, approximately 10 nm or less. After removing from the vacuum apparatus, the thin layer of carbon *plus* adsorbed protein or virus, in the form of single randomly dispersed particles, partly ordered 2-D arrays or 2-D crystals, should be floated on to the surface of the negative stain in a small petri dish. Depending on the thickness of the carbon film, it can be recovered from beneath the fluid surface on to the dull surface of bare 400 mesh copper grids or on to a holey carbon film over 200 or 300 mesh grids for a very thin floating carbon. Excess negative stain is then removed by carefully wiping the *underside* of the grid, and the grid allowed to air-dry. Note that with this technique the sample will be facing downwards when the grid is placed on to a filter paper. Thus, it is important for the negative stain to dry completely before this final positioning of the grid.

A minor modification of the above procedure, by using distilled water instead of negative stain at the final stage, enables the AM–PEG and any other sample salts or surfactant to be washed away and specimen grids to be directly blotted from beneath (the shiny side) prior to rapid plunge freezing for vitrification (see Section 7.1.1, and *Figure 8.10*).

The NS–CF procedure can also be performed on 20 µl droplets of negative stain or water within the micro-wells of a Teflon block. Approximately 3 mm square pieces of mica, with the four corners removed, should be cut from the larger carbon-coated mica with the spread sample, as shown in *Figure 3.13*. Because of surface tension, the insertion of the bare 400 mesh or glow-discharge treated holey-carbon grid requires that a small volume of stain or water be applied to the grid and almost completely shaken off on to a tissue paper, before inserting into the droplet *beneath* the tiny film of floating carbon. A small flexibly mounted or hand-held lens may usefully assist this manipulation. The individual grids produced this way can

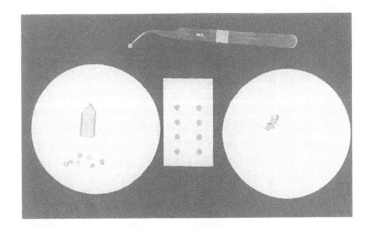

Figure 3.13: The use of wells in a Teflon block for the micro-version of the negative staining–carbon film technique, and for direct adsorption of protein on to a lipid layer already attached to a carbon film (Section 3.5.4).

The very small pieces (approx. 3 mm square) of carbon-coated mica, with corners removed, are inserted into water or negative stain droplets in the micro-wells. This releases the carbon film (as in *Figure 3.12*) and the mica settles to the bottom of the well. Insertion of a minimally water-coated bare grid beneath the floating carbon facilitates the removal of the tiny piece of floating carbon on to the surface of a grid, followed by removal of excess stain with a filter paper wedge and air-drying.

subsequently be washed with droplets of water or further negative stain solution, as described for the routine droplet procedure (Section 3.3.1).

3.5.2 2-D crystallization of proteins

As indicated above, one of the principal benefits to be gained from the NS–CF procedure is the production of 2-D (monomolecular layer) protein crystals for TEM study. Because of the importance of this potential of the NS–CF technique, a few further comments are appropriate. As is conventional with 3-D crystallization procedures, anyone wishing to use the NS–CF technique for 2-D crystallization must be prepared to investigate varying crystallization conditions in an attempt to narrow down and then routinely use those conditions that are more favourable for crystal production. Crystallization conditions have to be established independently for each virus or protein molecule.

Firstly, the importance of purification and stability of the protein or virus under study cannot be overestimated. Two-dimensional crystallization of protein samples that have been long-stored or frozen are unlikely to be successful, although there are exceptions. The presence of a high buffer and salt concentration in the protein sample is not desirable; 10 mM Tris-HCl buffer with 10–20 mM NaCl is acceptable, but best results often stem from distilled water-dialysed protein. The pH of the AM-PEG solution for 2-D crystallization can be selected only by trial and error. The isoelectric point (pI) of the protein is certainly of relevance as is the solubility of the protein close to its pI. Often, 3-D protein crystallization is successful at a pH slightly above the pI of the protein. If gross precipitation of protein is encountered at low pH it is natural to try a somewhat higher pH to prevent this, since protein aggregation will *always* interfere with crystallization. The M_r of the PEG used for the NS–CF procedure should usually be in the range 1000–10 000. The application of excess protein to the mica surface is not beneficial for the production of 2-D crystals, since once a monolayer of protein has formed, random molecular superimposition often occurs. Nevertheless, 3-D microcrystals can be produced by the NS–CF technique, particularly if PEG alone is mixed with the sample before spreading on mica (Harris *et al.*, 1992a, and see Chapter 5).

One feature that may be observed on electron micrographs from specimens produced by the NS–CF technique is the considerably varying depth of negative stain on the specimens, even from one edge of a 2-D crystal to the other. This uneven and/or partial-depth negative stain may not detract too seriously from the detail within the individual 2-D projection images, but the possibility for 3-D image reconstruction from a series of tilted images will be *severely* compromised. Thus, attempts to ensure that 2-D crystals are more completely covered by negative stain are desirable. This may be achieved by using the trehalose-containing negative stain solutions, which always tend to produce a thicker layer of stain, and/or by the inclusion of a low concentration (0.1–1.0 mM) of OG in the negative stain. Within the NS–CF technique, the carbon film should almost certainly be looked upon as a beneficial stabilizing support for 2-D crystals. But, if the inherent hydrophobicity and carbon–protein adsorption interferes with negative stain access or induces localized protein flattening, the preparation conditions cannot be considered optimal. The true magnitude of this problem remains to be quantified; in my opinion it is not excessive, but it cannot be ignored.

3.5.3 Cell cleavage on mica

The attachment of cells to any positively charged surface is a well known phenomenon, due primarily to available cell surface sialic acid carboxyl groups. A positive charge can readily be imparted to freshly cleaved mica by brief immersion in the strongly cationic dye Alcian blue (e.g. 0.1% w/v) or polylysine, followed by thorough washing in distilled water to remove unbound reagent. This approach has been developed as a variant of the NS–CF technique (Harris, 1991b), although earlier somewhat similar 'wet' cleavage procedures lead to the production of metal shadowed specimens. Once attached to the Alcian blue-coated mica, cell surface labelling can be performed and followed by highly efficient washing procedures, such as vertical immersion of the mica in phosphate-buffered saline (PBS) or other

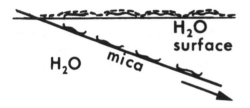

Figure 3.14: A schematic diagram showing cell wet cleavage on Alcian blue-treated mica (after Harris, 1991b).

Although the bound cells shown in this example are erythrocytes, almost any suspended cellular or purified organelle material can be attached to the mica and handled in a similar manner. This procedure can be performed using small petri dishes or Teflon block micro-wells (see *Figures 3.12* and *3.13*).

suitable medium, or simply flooding the mica surface from a pipette (see Nermut and Eason, 1989).

Procedure. The whole procedure can be performed on the usual small pieces of cleaved mica, as depicted in *Figure 3.14*. A cell suspension (e.g. red blood cells, platelets or cultured cells, etc.) should be applied to Alcian blue-coated mica and after a short time, during which some cells become attached, unbound cells are washed away by flooding on to a tissue paper with PBS or cell culture fluid. Brief stabilization of the bound cells can be performed using culture fluid (*without* fetal calf serum) containing 0.1% (v/v) glutaraldehyde. Higher concentrations of glutaraldehyde are not desirable, because complete fixation/cross-linkage of the cell and its contents is not required in this instance. (Cell surface immunolabelling can readily be introduced at this stage; see Chapter 4.1.)

Because the PBS salts, and other dissolved material in culture fluid interfere with the procedure, it is necessary to thoroughly wash the bound cells with a volatile buffer (e.g. 0.25 M ammonium acetate or ammonium carbonate) containing 20% (v/v) high purity glycerol. The pieces of mica are then placed within a vacuum coating apparatus which is allowed to evacuate to 1×10^{-5} Torr or better, for at least 2 h, to remove all volatile salts and ideally all the glycerol too. The exposed external surface of the cells is then coated with a thin layer of carbon. As with the original NS–CF procedure, the carbon film should then be removed from the mica using a small petri dish or micro-wells in a Teflon block (see *Figures 3.11* and *3.13*); in this instance it will take with it the adsorbed upper surface of the cells, released by the 'wet' cleavage process, leaving the lower surface and most of the cell contents attached to the positively charged mica. This step can be performed directly on a negative stain, or on water followed by droplet negative staining of the material attached to the lower surface of the carbon. For subsequent immunogold labelling procedures of the inner plasma membrane surface and/or any attached membranous or cytoskeletal material, it will be necessary to use nickel or gold EM grids (Section 4.1.2).

The wide possibilities of this cell cleavage–negative staining technique have not been fully explored or exploited, but I believe that this approach has considerable potential, particularly in the areas of cell surface dynamics and cell surface antigen, receptor and cytoskeleton labelling. Because access is available to both the inner and outer surface of the plasma membrane at different stages of the procedure and to cytoskeletal elements and cytomembranes, comparative and multiple labelling studies are possible.

This procedure, without Alcian blue treatment of the mica, can indeed be used for any particulate biological suspensions that contain a high concentration of EM-incompatible solute such as glycerol, sucrose, urea or high salts (Harris, 1991b). Washing the mica with a buffer containing volatile salts such as ammonium acetate or ammonium carbonate removes these solutes and still leaves a sufficient quantity of loosely attached protein, which will then be adsorb more strongly to the freshly evaporated carbon, for subsequent negative staining.

3.5.4 Lipid monolayers and multilayers on mica

When a solution of lipid in an organic solvent, such as ethanol, methanol or chloroform–methanol is spread on freshly cleaved mica and excess fluid removed, drying occurs rapidly, leaving the crystalline/liquid-crystalline lipid on the mica surface. Depending upon the concentration of the lipid solution, the mica may be partly or completely covered with a lipid monolayer, bilayers and multi-bilayer crystals. Similarly, aqueous suspensions of lipid microcrystals or liposomes, even containing 10–20% (v/v) organic solvent can be spread upon a mica surface, giving an even spread of the particulate lipid material. In both cases the dried material on the mica is suitable for the production of platinum–carbon shadowed specimens or negatively stained specimens by the NS–CF procedure (Section 3.5.1). These approaches have been successfully used to study the multi-bilayer composition of pure cholesterol (Harris, 1988).

One very significant extension to this approach, as already indicated in Section 3.5.3, is the use of the biological material adsorbed to a floating film of carbon as an immobilized substrate for immunological, affinity and biochemical reactions. This procedure can most conveniently be performed using Teflon plated micro-wells (*Figure 3.11*). Thus, adsorbed single-layer erythrocyte membranes can be subjected to enzymic digestion of the membrane skeleton, in the knowledge that the outer surface will remain attached to the floating carbon. Access to the inner surface of the plasma membrane can then be obtained and sterically hidden antigenic epitopes may be made accessible (Nermut, 1982a). With carbon-adsorbed lipid monolayers and multi-bilayers it is possible to study the interaction of lipid-specific toxins and other substances that are known to require lipids for their functional activity, such as ion channels, lipases and phospholipases (for antibody and immunogold labelling, see Section 4.1).

3.5.5 Dynamic and biochemical aspects of the NS–CF technique

The early observations of Bob Horne and his colleagues indicated that during the NS–CF technique it was possible to obtain dissociation and reassembly of viral components, depending upon the conditions used (reviewed by Horne, 1986; Harris and Horne, 1994). It appears that the action of ammonium molybdate is of prime importance in these effects. Recent studies on molluscan haemocyanin (Harris *et al.*, 1995b) have indicated that during the NS–CF procedure and *in solution*, ammonium molybdate will specifically dissociate keyhole limpet haemocyanin type 2 (KLH2) in a progressive and pH-dependent manner, over the range pH 5.5–6.5, with complete dissociation close to the lower pH value. Remarkably, over this pH range keyhole limpet haemocyanin type 1 (KLH1) does not dissociate. This biochemical treatment was, in retrospect, in full agreement with the fact that the NS–CF technique was able to create 2-D crystals from the KLH1 didecamer, and that even at neutral pH the KLH2 dissociated into decamers (Harris *et al.*, 1992b). More recent investigations using the NS–CF technique have shown that the helical tubular polymer of KLH1, formed by reassociation of subunits of this molecule produced at pH 9.6, will convert into a stacked-disc/parallel tube form and indeed the more usual didecameric form of the molecule (see Section 5.4). These examples serve to emphasize further the interesting possibilities of the NS–CF technique within the sphere of biochemical interactions which fall within the electron microscopical concept of macromolecular *microdissection* and structural reorganization. These aspects should, perhaps, be considered separately from the principal feature of mica-spreading, where dynamic molecular interactions involving forces at both the fluid–mica surface and the fluid–air interface are likely to be required for the initiation and progression of the 2-D and even 3-D crystallization process (Harris and Holzenburg, 1995).

References

Aebi U, Pollard TD. (1987) A glow discharge unit to render electron microscope grids and other surfaces hydrophilic. *J. Electr. Microsc. Technique* **7**, 29–33.

Akey CW, Edelstein SJ. (1983) Equivalence of the projected structure of thin catalase crystals preserved for electron microscopy by negative stain, glucose or embedding in the presence of tannic acid. *J. Mol. Biol.* **163**, 575–612.

Anderson TF. (1966) Electron microscopy of microorganisms. In *Physical Techniques in Biological Research* (ed. AW Pollister) Vol. III, Part A. Academic Press, New York, pp. 319–387.

Asturias FJ, Kornberg RD. (1995) A novel method for transfer of two-dimensional crystals from the air/water interface to specimen grids. *J. Struct. Biol.* **114**, 60–66.

Barnakov NA. (1994) Sequential treatment by phosphotungstic acid and uranyl acetate enhances the adherence of lipid membranes and membrane proteins to hydrophobic EM grids. *J. Microsc.* **175**, 171–174.

Benedetti EL, Emmelot P. (1968) Hexagonal array of subunits in tight junctions separated from rat liver plasma membranes. *J. Cell Biol.* **38**, 15–24.

Bremner A, Hann C, Engel A, Baumeister W, Aebi A. (1992) Has negative staining a future in biomolecular electron microscopy? *Ultramicroscopy* **46**, 85–111.

Brisson A, Olofsson A, Ringler P, Schmutz M, Stoylova S. (1994) Two-dimensional crystallization of proteins on planar lipid films and structure determination by electron crystallography. *Biol. Cell* **80**, 221–228.

Cyrklaff M, Roos N, Gross H, Dubochet J. (1994) Particle–surface interaction in thin vitrified films for cryo-electron microscopy. *J. Microsc.* **175**, 135–142.

Dubochet J, Ducommun M, Zollinger M, Kellenberger E. (1971) A new preparation method for dark-field electron microscopy of biomacromolecules. *J. Ultrastruct. Res.* **35**, 147–167.

Frank J. (1989) Image analysis of single molecules. *Electr. Microsc. Rev.* **2**, 53–74.

Fujiyama Y, Yokoyama K, Yoshida M, Wakabayashi T. (1990) Electron microscopy of the reconstituted complexes of the F_1–ATPase with various subunit constitution revealed the location of the subunit in the central cavity of the molecule. *FEBS Lett.* **271**, 111–115.

Fukami A, Adachi K. (1965) A new method of preparation of a self-perforated micro-plastic grid and its applications. *J. Electr. Microsc. (Japan)* **14**, 112–118.

Ghiretti-Magaldi A, Zanotti G, Togon G, Mezzalasalama V. (1985) The molecular architecture of the extracellular hemoglobin of *Ophelia bicornis*. *Biochim. Biophys. Acta* **829**, 144–149.

Gregory DW, Pirie BJS. (1973) Wetting agents for biological electron microscopy. 1. General considerations and negative staining. *J. Microsc.* **99**, 251–265.

Hainfeld JF, Safer D, Wall JS, Simon M, Lin B, Powell RD. (1994) Methylamine vanadate (Nanovan) negative stain. In *Proceedings of the 52nd Annual Meeting of MSA, New Orleans* (eds GW Baily, AJ Garratt-Reed). San Francisco Press, San Francisco, pp. 132–133.

Harris JR. (1982) The production of paracrystalline two-dimensional monolayers of purified protein molecules. *Micron* **13**, 169–184.

Harris JR. (1986a) A negative staining study of aqueous suspensions of sphingomyelin. *Micron Microsc. Acta* **17**, 175–200.

Harris JR. (1986b) A negative staining study of natural and synthetic L-α-lysophospha-tidylcholine micelles, macromolecular aggregates and crystals. *Micron Microsc. Acta* **17**, 289–305.

Harris JR. (1988) Electron microscopy of cholesterol. *Micron Microsc. Acta* **19**, 19–32.

Harris JR. (1991a) The negative staining carbon film procedure: technical considera-tions and a survey of macromolecular applications. *Micron Microsc. Acta* **22**, 341–359.

Harris JR. (1991b) Negative staining–carbon film techniques: new cellular and molecular applications. *J. Electr. Microsc. Technique* **18**, 269–276.

Harris JR, Agutter PS. (1970) A negative staining study of human erythrocyte ghosts and rat liver nuclear membranes. *J. Ultrastruct. Res.* **33**, 219–232.

Harris JR, Holzenburg A. (1989) Transmission electron microscopy of negatively stained human erythrocyte catalase. *Micron Microsc. Acta* **20**, 223–238.

Harris JR, Holzenburg A. (1995) Human erythrocyte catalase: 2-D crystal nucleation and multiple 2-D crystal forms. *J. Struct. Biol.* **115**, 102–112.

Harris JR, Horne RW. (1991) Negative staining. In *Electron Microscopy in Biology: a Practical Approach* (ed. JR Harris). IRL Press, Oxford, pp. 203–228.

Harris JR, Horne RW. (1994) Negative staining: a brief assessment of current technical benefits, limitations and future possibilities. *Micron* **25**, 5–13.

Harris JR, Kerr J. (1976) Contrast enhancement of negatively stained macromolecules and biological membranes by single side band phase contrast interference. *J. Microsc.* **108**, 51–59.

Harris JR, Pfeifer G, Pühler G, Baumeister W. (1992a) Production of 3-D microcrystals from *Thermoplasma acidophilum* multicatalytic proteinase/proteosome by the negative staining–carbon film techniques. In *Electron Microscopy*, Vol. 1. EUREM 92, Granada, Spain, pp. 383–384.

Harris JR, Cejka Z, Wegener-Strake A, Gebauer W, Markl J. (1992b) Two-dimensional crystallization, transmission electron microscopy and image processing of keyhole limpet hemocyanin. *Micron Microsc. Acta* **23**, 278–301.

Harris JR, Depoix F, Urich K. (1995a) The structure of gas-filled n-butyl-2-cyanoacrylate (BCA) polymer particles. *Micron* **26**, 103–111.

Harris JR, Gebauer W, Söhngen SM, Markl J. (1995b) Keyhole limpet hemocyanin (KLH): purification of intact KLH1 through selective dissociation of KLH2. *Micron* **26**, 201–212.

Harris JR, Gebauer W, Markl J. (1995c) Keyhole limpet hemocyanin (KLH): negative staining in the presence of trehalose. *Micron* **26**, 25–33.

Harris JR, Zahn R, Plückthun A. (1995d) Electron microscopy of the GroEL–GroES filament. *J. Struct. Biol.* **115**, 68–77.

Hayat MA, Miller SE. (1990) *Negative Staining*. McGraw-Hill, New York.

Horne RW. (1977) Optical diffraction analysis of periodically repeating biological structures. In *Analytical and Quantitative Methods in Microscopy* (eds GA Meek, HY Elder). Cambridge University Press, Cambridge, pp. 29–53.

Horne RW. (1986) Electron microscopy of crystalline arrays of adenoviruses and their components. In *Electron Microscopy of Proteins* (eds JR Harris, RW Horne). Academic Press, London, pp. 71–101.

Horne RW, Pasquali-Ronchetti I. (1974) A negative staining–carbon film technique for studying viruses in the electron microscope. I. Preparation procedures for examining isosahedral and filamentous viruses. *J. Ultrastruct. Res.* **47**, 361–383.

Horne RW, Wildy P. (1979) An historical account of the development and applications of the negative staining technique to the electron microscopy of viruses. *J. Microsc.* **117**, 103–122.

Jahn W. (1995) Easily prepared holey films for use in cryo-electron microscopy. *J. Microsc.* **179**, 333–334.

Jap BK, Zulauf M, Scheybani T, Hefti A, Baumeister W, Aebi U, Engel A. (1992) 2D crystallization: from art to science. *Ultramicroscopy* **46**, 45–84.

Johnson RPC, Gregory DW. (1993) Viruses accumulate spontaneously near droplet surfaces: a method to concentrate viruses for electron microscopy. *J. Microsc.* **171**, 125–136.

Kiselev NA, Sherman MB, Tsuprun VC. (1990) Negative staining of proteins. *Electr. Microsc. Rev.* **3**, 43–72.

Kodama M, Miyata T, Takaichi Y. (1993) Calorimetric investigations of phase transitions of sonicated vesicles of dimyristoylphosphatidylcholine. *Biochim. Biophys. Acta* **1169**, 90–97.

Kubalek EW, Kornberg RD, Darst SA (1991) Improved transfer of two-dimensional crystals from the air/water interface to specimen support grids for high-resolution analysis by electron microscopy. *Ultramicroscopy* **35**, 295–304.

Kühlbrandt W. (1992) Two-dimensional crystallization of membrane proteins. *Quart. Rev. Biophys.* **25**, 1–49.

Lake JA. (1981) Protein synthesis in prokaryotes and eukaryotes: the structural basis. In *Electron Microscopy of Protein* (ed. JR Harris) Vol. 1. Academic Press, London, pp. 167–195.

Massover W. (1975) Borates as agents for negative staining. In *Proceedings of the 33rd Annual Meeting of the Electron Microscopy Society of America* (ed. GW Bailey). Claitor's, Baton Rouge, LA, pp. 630–631.

Mellema JE, van Bruggen EFJ, Gruber M. (1967) Uranyl oxalate as a negative stain for electron microscopy of proteins. *Biochim. Biophys. Acta* **140**, 180–182.

Misel D. (1978) Image analysis, enhancement and interpretation. In *Practical Methods in Electron Microscopy* (ed. AM Glauert) Vol. 7. North-Holland, Amsterdam, pp. 48–53.

Munn EA. (1968) On the structure of mitochondria and the value of ammonium molybdate as a negative stain for osmotically sensitive structures. *J. Ultrastruct. Res.* **25**, 363–380.

Muscatello U, Horne RW. (1968) Effect of the tonicity of some negative staining solutions on the elementary structure of membrane-bounded systems. *J. Ultrastruct. Res.* **25**, 73–83.

Namork E, Johansen BV. (1982) Surface activation of carbon film supports for biological electron microscopy. *Ultramicroscopy* **7**, 321–330.

Nermut MV. (1982a) The cell monolayer technique in membrane research: a review. *Eur. J. Cell Biol.* **28**, 160–172.

Nermut MV. (1982b) Advanced methods in electron microscopy of viruses. In *New Developments in Practical Virology* (ed. CR Howard). Alan R. Liss, New York, pp. 1–58.

Nermut MV. (1991) Unorthodox methods of negative staining. *Micron Microsc. Acta* **22**, 327–339.

Nermut MV, Eason P. (1989) Cryotechniques in macromolecular research. *Scann. Microsc.* Suppl. 3, 213–225.

Nogales E, Wolf SG, Zhang SX, Downing KH. (1995) Preservation of 2-D crystals of tubulin for electron microscopy. *J. Struct. Biol.* **115**, 199–208.

Oliver RM. (1973) Negative stain electron microscopy of protein macromolecules. In *Methods in Enzymology* (eds CHW Hirs, S Timasheff) Vol. XXVII. Academic Press, New York, pp. 616–672.

Peters J-M, Harris JR, Lustig A, Müller S, Engel A, Volker S, Franke WW. (1992) The ubiquitous Mg^{2+}–ATPase complex: a structural study. *J. Mol. Biol.* **223**, 557–571.

Phillips ML, Schumaker VN. (1989) Conformation of apolipoprotein B after lipid extraction of low density lipoproteins attached to an electron microscope grid. *J. Lipid Res.* **30**, 415–422.

Schultz P, Celia H, Riva M, Darst SA, Colin P, Kornberg RD, Sentenac A, Oudet P. (1990) Structural study of the yeast RNA polymerase A. Electron microscopy of lipid-bound molecules and two-dimensional crystals. *J. Mol. Biol.* **216**, 353–362.

Schwartz H, Riede L, Sonntag I, Henning U. (1983) Degrees of relatedness of T-even type *E. coli* phages using different or the same receptors and topology of serological cross-reacting sites. *EMBO J.* **2**, 375–380.

Somerville J, Scheer U. (1987) *Electron Microscopy in Biology: a Practical Approach.* IRL Press, Oxford.

Spiess E, Zimmermann H-P, Lunsdorf H. (1987) Negative staining of protein molecules and filaments. In *Electron Microscopy in Biology: a Practical Approach* (eds J Sommerville, U Scheer). IRL Press, Oxford, pp. 147–166.

Steven AC, Smith PR, Horne RW. (1978) Capsid fine structure of cowpea chlorotic mottle virus: from a computer analysis of negatively stained virus arrays. *J. Ultrastruct. Res.* **64**, 63–73.

Tranum-Jensen J. (1988) Electron microscopy: assays involving negative staining. In *Methods in Enzymology* (eds CHW Hirs, S Timasheff) Vol. 165. Academic Press, New York, pp. 357–374.

Unwin PNT. (1974) Electron microscopy of the stacked disk aggregate of TMV protein. II. The influence of electron irradiation on the stain distribution. *J. Mol. Biol.* **87**, 657–670.

Valentine RC, Shapiro BM, Stadtman ER. (1968) Regulation of glutamine synthetase. XII. Electron microscopy of the enzyme from *Escherichia coli*. *Biochemistry* **7**, 2143–2152.

Wang DN, Kühlbrandt W. (1991) High-resolution electron crystallography of light-harvesting chlorphyl a/b-protein complex in three different media. *J. Mol. Biol.* **217**, 691–699.

Wells B, Horne RW, Shaw PJ. (1981) The formation of two-dimensional arrays of isometric plant viruses in the presence of polyethylene glycol. *Micron* **12**, 37–45.

4 Negative Staining: Some Specialized Approaches and Problems

4.1 Immunonegative staining

4.1.1 Direct antibody labelling

The principal aim of antibody labelling of biological materials is to define the surface localization of antigenic epitopes on cellular structures and isolated components such as ribosomes and macromolecules. This approach has successfully been combined with negative staining for many years, as reviewed by Boisset and Lamy (1991), Harris (1996) and Lünsdorf and Tiedge (1992).

Immunonegative staining was initially used by plant, animal and human virologists (Carrascosa, 1988; Roberts, 1986), who were able to show that purified viruses could be coated with a layer of antibody and that viruses would attach to carbon support films already containing adsorbed antibody. Although this diagnostic approach has been largely superseded by other quantitative immunological assays (e.g. ELISA) it remains extremely powerful for structural studies. This has been increasingly demonstrated with the availability of monoclonal antibodies and peptide-specific polyclonal antibodies against known amino acid sequences within proteins. Studies can be performed by defining the location of single IgG, Fab and Fab' molecules attached to an epitope or by the interpretation of the IgG and Fab linkage pattern produced within small immune complexes. In general, the latter is somewhat easier and almost certainly more accurate, but the former approach using Fab', has a potentially higher resolution for epitope location, if used with care.

Details of the immunological aspects (i.e. antibody production and immune complex formation) will not be given here, but it should be said that it is often desirable to have purified antigen available and if possible purified and concentrated monoclonal IgG. However, separation of immune complexes by gel filtration chromatography from unreacted antigen and antibody immediately prior to specimen preparation does, to a large extent, overcome the need for a very high purity antigen, and monoclonal antibody in hybridoma cell culture supernatants can also often be used without purification.

Whatever the procedure used for producing single molecules with at-tached antibody or antigen–antibody complexes, the preparation of nega-tively stained specimens is essentially as described for the single droplet procedure (Section 3.3.1). Uranyl acetate, uranyl formate, sodium phosphotungstate and ammonium molybdate can be successfully used (Cejka *et al.*, 1993; Harris *et al.*, 1993b; Kopp *et al.*, 1993, 1995; Martin *et al.*, 1994; see examples in Chapter 5). For immune complexes prepared from high molecular weight proteins it may be found that the inclusion of trehalose is helpful, as this sugar maintains the 3-D conformation of the immune complexes in a superior manner, by reducing sample flattening on the carbon support film. It is desirable that specimens of immune com-plexes should be prepared soon after the elution of the material from a gel filtration column, in order to prevent further immune-aggregation or pos-sibly dissociation of the complexes. As the chromatography will usually be performed in the presence of a buffered saline solution, it is always neces-sary to employ several droplet washes with distilled water, after attach-ment of the immune complexes to the carbon support film. If the protein concentration is rather low, adsorption times of 2 min or longer can be employed, but in this situation it is likely that uneven spreading of the negative stain will occur around well-spaced immune complexes.

Interpretation of the binding site of a single IgG, Fab or Fab' on the surface of a protein or biological structure such as bacteriophage tail fibre (Schwartz *et al.*, 1983), may provide an accurate definition of the epitope location. Single particle image processing can also provide some further enhancement of the interpretation (Lamy *et al.*, 1990), but is unlikely to produce a clear result from visually doubtful images. The specific location of a number of identical epitopes on the surface of a multi-subunit homo- or hetero-oligomeric protein molecule may be firmly indicated by the steric pattern adopted within the immune complex (Kopp *et al.*, 1993, 1995). This pattern may only be apparent with relatively small groups of mol-ecules, since the creation of large 3-D complexes in solution can give rise to considerable overlapping of molecules on the negatively stained speci-men (see Harris, 1996). It should, perhaps, be emphasized that studies on immune complexes can also be performed successfully with unstained fro-zen-hydrated specimens (Boisset *et al.*, 1995) and that in these instances the use of computer processing may very often be indispensable, because of the low image contrast.

4.1.2 Immunogold labelling

If the labelling resolution sought is at the level of cellular structure, such as cytoskeletal intermediate filaments or microtubules, rather than the more precise location of an epitope on an isolated macromolecule or virus, then direct or indirect immunogold labelling in conjunction with negative staining may provide the information required. This technical approach has been thoroughly documented by Hyatt (1991). Colloidal gold sols and gold conjugates, with a range of IgGs, protein A and protein G, and streptavidin and avidin are widely available commercially. For negative staining, useful gold particle sizes range from 2 nm to 10 nm. For direct labelling, conjugates need to be individually prepared and stabilized using purified IgG, Fab or Fab' (Oliver, 1994)), but for indirect/secondary labelling the commercially available conjugated products will cover most eventualities. The technique to be described below can also be utilized for the investigation of lectin–membrane and toxin–membrane interactions, which may involve a protein or glycoprotein receptor as the ligand or a membrane lipid, such as cholesterol. In these latter instances it may be possible to utilize direct labelling with colloidal gold conjugated to lectin or toxin, or to avidin/streptavidin for use with biotinylated protein, such as streptolysin O (see Chapter 5).

For viruses, macromolecules, isolated membranes and reconstituted lipid–protein systems it is possible to perform direct labelling with immunogold 'in solution', as long as the addition of an excessive quantity of label is avoided. Negatively stained specimen can then be prepared by the routine droplet procedure (Section 3.3.1). The alternative approach is firstly to attach the material under study to a carbon–plastic support film, followed by a sequential series of 'on-grid' labelling and washing steps, and finally negative staining. Cellular cytoskeletal labelling can be performed on cell monolayers grown on grids, following (partial) removal of membranes by neutral surfactant extraction (Hyatt, 1991). Because of the reactivity of copper in saline-containing solutions during the prolonged incubations, it is necessary to use nickel or gold grids for 'on-grid' immunogold labelling. A generalized procedure, using erythrocyte membranes as an example, will be given below, which can be adapted readily to the specific requirements of any particular experimental situation and biological material.

Procedure. Sample material (e.g. haemoglobin-free erythrocyte 'ghosts', purified viral particles or macromolecules) at a protein concentration of approximately 0.5 mg ml^{-1} should be attached to a glow-discharge treated carbon–plastic support film, such that no overlapping of the individual membranes or particles occurs. Free membranes can be washed away with a 20 µl droplet of PBS, and the grid floated (sample down) on a droplet of primary antibody or for direct immunogold labelling with antibody (IgG/Fab/Fab')-gold, at an appropriate dilution and length of time (usually 30 or 60 min) at room temperature. During this time the droplet plus grid should be covered to reduce evaporation. Free antibody should then be washed away with a series of PBS-droplet washes. Inclusion of Triton-X100 or other neutral surfactant will reduce non-specific background, but may destroy the membranes; bovine serum albumin or bovine calf serum can be used instead. For labelling protein alone, and also non-membranous viruses, neutral surfactant can safely be included, indeed its continued inclusion is beneficial for labelling of cytoskeletal structures *in situ*.

For secondary labelling gold–anti-goat/rabbit/mouse/guinea pig or gold–protein A/protein G labelling is performed in the same way after incubation in the primary antibody alone, followed by thorough PBS and finally water washing, before droplet-negative staining. With erythrocyte ghosts, negative staining should be performed with neutral 2% ammonium molybdate or 1% uranyl acetate; sodium phosphotungstate produces marked membranous changes and should not be used in this instance, although it may be perfectly suitable for proteins and viruses. All possible controls should be included to ensure that the antibody labelling reactions being used are truly specific.

Extensive individual variations should be employed for each system under investigation until a satisfactory and reproducible protocol is established. Clear positive labelling with colloidal gold will usually be immediately convincing and of value, whereas dubious/borderline labelling with a high background is of little value and should always be acknowledged as such. Extensive documentation of colloidal gold labelling techniques can be found within the series of books edited by M.A. Hayat (1989–1991).

4.1.3 Nanogold particles and metal atom cluster labels

A recent important technical innovation in the field of macromolecular labelling has been the development of gold particles, of diameter 1.4 nm (Nanogold™ [a]). These gold compound particles can be chemically conjugated to purified IgG molecules and Fab' fragments via the hinge -SH group for immunolabelling (Watts *et al.*, 1990), and they can also be used as direct labels to locate available -SH or lipoyl groups on macromolecules. It has been maintained by Hainfeld *et al.* (1994) that the low contrast negative stain methylamine vanadate is most suitable for negative staining after 1.4 nm Nanogold™/gold compound and 0.8 nm undecagold/gold cluster labelling, since it does not mask the metal atoms. The use of sodium tetraborate, lower than usual concentrations of the more widely used negative stains or more promisingly negative stain–trehalose combinations might well achieve the same result, since it is possible to readily adjust the contrast by reducing the negative stain concentration whilst maintaining the carbohydrate at 1% w/v.

The 0.8 nm diameter gold cluster undecagold (Au^{11}) product also possesses considerable potential for even higher resolution immunolabelling and direct labelling of thiol, carbohydrates and primary amines. The location of -SH bound gold clusters has also been successfully achieved by cryoelectron microscopy (Boisset *et al.*, 1992; Lambert *et al.*, 1994). Scanning transmission electron microscopy (STEM) also provides a useful ap-

---

[a] Nanogold™ is the trade name of the 1.4 nm gold compound product from Nanoprobes Inc., Stoney Brook, NY, USA.

proach for the location of gold clusters (Yang *et al.*, 1994). Instead of presenting protocols for the use of the commercially available Nanogold™ and undecagold the reader is referred to the useful product information from Nanoprobes Inc. and the review by Koeck and Leonard (1996).

4.2 Freezing and negative staining

Two freezing variants of negative staining were both developed by Nermut and his colleagues, with emphasis upon studies dealing with viruses and membranes (Nermut, 1991; Nermut and Frank, 1971). These somewhat more complex negative stain approaches hold some considerable potential for the resolution of fine structure, by preventing particle collapse.

4.2.1 Freeze-dry negative staining

Procedure. A monolayer of virus particles (or protein molecules) should be prepared on carbon-coated grids by either: (a) spreading the virus suspension on the surface of 0.25 M ammonium acetate or distilled water in a petri dish using the established 'Kleinschmidt technique' (Kleinschmidt, 1968), then with fine forceps, placing glow-discharge treated carbon-treated grids on the fluid surface and removing; or (b) by floating glow-discharge treated carbon-coated grids on a 20 µl droplet of virus suspension, with removal of unadsorbed virus by repeated washing on three or four drops of ammonium acetate or water (as in Section 3.3.1). Grids should then be placed on a 20 µl droplet of negative stain solution for 10–20 sec, and the excess liquid removed by a filter paper. *Before* they have a chance to air-dry, the grids are then plunged rapidly into liquid cryogen (nitrogen, propane or ethane). While still frozen the grids should then be transferred on to a precooled specimen stage at approximately –100°C of a conventional freeze-etch unit (e.g. BAL-TEC or Edwards High Vacuum) and the equipment pumped down to a vacuum of 10^{-5} Torr. To induce freeze-drying, the temperature of the specimen stage is then adjusted to –80 to –85°C, while the temperature of a cool trap within the chamber is kept at approximately –150°C. The specimen should freeze-dry in 20–30 min, with a visible white powder appearing on the grid surface. After warming the specimen stage to 30°C, specimen grids can be removed and the white powder carefully blown off the grids with a gentle stream of dry air or nitrogen gas. Grids are then ready for immediate observation in the electron microscope.

In general, the freeze-dry negative staining procedure will reveal surface features of virus particles, which are in a well preserved state, with no flattening (see *Figures 5.26* and *5.27*).

4.2.2 Freeze–fracture negative staining

Procedure. Prepare a monolayer of virus particles, membranes or cells (e.g. red blood cells, suspension culture cells or bacteria) in a petri dish and transfer to the surface of freshly cleaved mica which has been rendered positively charged with polylysine or Alcian blue. Make a sandwich of the biological sample with a thin copper plate (or a second piece of uncleaved mica) and plunge the two into a liquid cryogen (nitrogen slush or liquid ethane is recommended). While still in the cryogen, separate the mica from the copper, to produce fractured particles. Remove and warm up the positively charged piece of mica and without delay: (a) immerse beneath the surface of a suitable buffer, or (b) deposit a small drop of buffer on to the mica and place glow-discharge treated carbon-coated grids

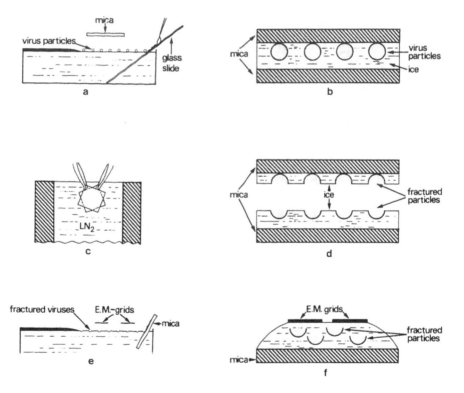

Figure 4.1: A diagrammatic representation of the freeze–fracture negative staining procedure.
A monolayer of virus particles is formed by water-spreading and is picked up on to a small piece of cleaved mica (a); the mica is sandwiched with another (uncleaved) piece of mica or a thin copper plate (b) and plunged into liquid cryogen (c). The sample is fractured by separating the mica pieces under cryogen (c, d). After thawing the mica is slowly submerged in water and the virus particles are picked up by touching the surface with a carbon-coated grid (e). Alternatively, grids can be placed on the mica for a few minutes to allow adsorption to take place (f). Cell or membrane monolayers on positively charged glass or mica can be processed in a similar manner to viruses, but spreading after freeze–fracture as in (e) is not recommended. (Procedure following Nermut, 1991.)

on top (see *Figure 4.1*). Remove the grids, wash with water droplets and then negatively stain, as in Section 3.3.1.

4.3 Problems and artefacts of negative staining

As mentioned previously, negative staining with the slightly acidic uranyl acetate (approx. pH 4.5) often produces some element of positive staining, due to direct interaction of the uranyl cation with available negatively charged phosphate and carboxyl groups. Positive staining can be observed most clearly if the sample is subsequently water-washed and then studied *without* negative stain. This positive staining has, however, been found to stabilize many proteins and viruses and can thereby reduce structural collapse during air-drying, prior to metal shadowing. For routine specimens negatively stained with uranyl acetate there does not appear to be a problem, since it becomes essentially impossible to distinguish between uranyl acetate bound to the surface negatively charged groups of a protein and that which is freely surrounding and permeating the aqueous spaces. For higher resolution studies some discrimination between positively bound stain and stain that is simply permeating aqueous spaces might be desirable (Woodcock and Baumeister, 1990).

The situation with the negative stains that produce direct structural changes, such as viral disruption, protein dissociation and membrane destabilization is rather different. Such changes may sometimes be used constructively, even as biochemical dissociation systems once they are clearly recognized, understood and controlled, whereas when they occur spuriously or inadvertently without the prior knowledge of the observer, they present a considerable danger.

Perhaps one of the major hazards of negative staining is the situation where the negative stain does not spread evenly over the carbon support film. This is most likely when the protein concentration on the carbon is too low, or because glow-discharge treatment was not used or was inadequate. Small and large hydrophobic patches on the carbon may exclude the stain and should always be carefully distinguished from the biological material under investigation. There are a number of published cases where such 'background' negative staining defects have been claimed as genuine 'structural features', when in reality they are often meaningless! Similarly, a hazard of over-activation of the carbon by excessive glow discharge can lead to selective orientation of protein molecules, which results in a limited or even false interpretation of the true structure. For instance, the cylindrical 20S proteasome has been incorrectly described as 'ring-like', because it was imaged *only* from the 'end-on' orientation in negatively stained specimens (cf. Kleinschmidt *et al.*, 1983; Tanaka and Ichihara, 1990).

When the negative stain depth is less than the thickness of the biological material under investigation it is clear that incomplete and inaccurate projection imaging will be achieved (Harris *et al.*, 1993a). This situation may not be too limiting in the first instance, as the overall size and shape of a protein molecule may be correct, but when it comes to defining the size and shape of superimposed subunits severe errors will be introduced and such images will not contain the necessary information to enable 3-D reconstructions to be performed. However, using the known limitation of negative stain depth, Nermut and Perkins (1979) were able to successfully define a *very significant* structural difference between the terminal region and base of the adenovirus hexon (see also Chapter 5*)*.

With negative stains containing glucose or trehalose, the increased depth of embedding stain brings considerable benefits, but also its own special difficulties. The liability of carbohydrate in the electron beam is the first undesirable feature although it can be countered quite successfully by maintaining the minimum possible electron dose and by specimen cooling (e.g. to –170°C). However, regions of excessively thick negative stain containing carbohydrate are still likely to show rapid deterioration in the electron beam, with pronounced 'bubbling' and loss of structural detail within the biological material (*Figure 4.2*). This unacceptable situation appears to be somewhat worse with glucose than trehalose. The

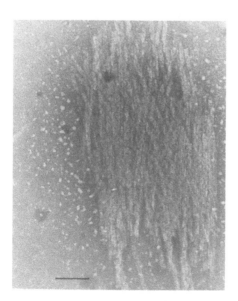

Figure 4.2: A paracrystalline bundle of helical tubules produced by reassociation of KLH1, embedded in a thick layer of 5% ammonium molybdate–1% trehalose.

The negative stain shows considerable 'bubbling' due to electron beam damage. Negatively stained specimens such as this need to be studied under strict low electron dose conditions. Somewhat shallower regions of negative stain do not exhibit such marked beam damage (see Figure 5.22) The scale bar indicates 200 nm.

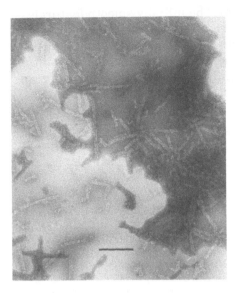

Figure 4.3: Double-sided negative staining.
  An example of double-sided negative staining, with somewhat deeper negative stain showing at the RHS. Such specimens should usually be discarded. The scale bar indicates 200 nm.

granularity/microcrystallinity of uranyl acetate–trehalose increases more markedly and rapidly than ammonium molybdate–trehalose during electron irradiation.

Partial breakage of the carbon support film during the droplet negative staining procedure will often allow some sample and negative stain to spread on to the opposite side of the carbon. This can result in a double-sided specimen with overlapping of the sample, often with a difference in stain depth on the two carbon surfaces, since only one surface of the carbon has received the beneficial effects of glow discharge (*Figure 4.3*). This unsatisfactory situation is usually obvious and can be readily avoided at the specimen preparation stage, when fluid spread on to the opposite carbon surface can be seen and the specimen grid discarded. The appearance of spurious salt crystals within the negative stain is a common problem during the production of negatively stained specimens. Whilst usually due to inadequate water washing, it can be due to specimen buffer salts that have inadvertently spread to both sides of the carbon, but have been washed away predominantly on one side, some salts remaining to diffuse into the negative stain layer before it dries. An improved droplet washing technique, with more careful positioning of the forceps only at the very edge of the grid and avoidance of 'wicking' into the tips of the forceps, usually cures this difficulty. Anomalous drying of ammonium molybdate may be seen when the solution is in excess of pH 7.0. This results in obvious electron-dense crystals within the surrounding less dense amorphous negative stain, an effect that may tend to become worse as the stain solution ages (AM alone, but *not* AM–trehalose). The initially smooth/featureless vitreous nature of air-dried trehalose–negative stain mixtures on the specimen grid appears to undergo some form of 'crystallization', leading to undesirable specimen alteration over a period of several weeks or months. This change is readily detected and easily avoided if specimens are studied reasonably rapidly following their preparation (i.e. within c. 4 weeks).

References

Boisset N, Lamy J. (1991) Immunoelectron microscopy and image processing for epitope mapping. In *Methods in Enzymology* (ed. JJ Langone) Vol. 203. Academic Press, New York, pp. 274–295.

Boisset N, Grassucci R, Penczek P, Delain E, Pochon F, Frank J, Lamy JN. (1992) Three-dimensional reconstruction of a complex of human α_2-macroglobulin with monomaleimido Nanogold (Au1.4nm) embedded in ice. *J. Struct. Biol.* **109**, 39–45.

Boisset N, Penczek P, Taveau J-C, Lamy J, Frank J, Lamy J. (1995) Three-dimensional reconstruction of *Androctonus australis* hemocyanin labeled with a monoclonal Fab fragment. *J. Struct. Biol.* **115**, 16–29.

Carrascosa JL. (1988) Immunoelectronmicroscopical studies on viruses. *Electr. Microsc. Rev.* **1**, 1–16.

Cejka Z, Gould-Kostka J, Burns D, Kessel M. (1993) Localization of the binding site of an antibody affecting ATPase activity of chaperonin cpn60 from *Bordetella pertussis*. *J. Struct. Biol.* **111**, 34–38.

Hainfeld JF, Safer D, Wall JS, Simon M, Lin B, Powell RD. (1994) Methyl vanadate (Nanovan) negative stain. In *Proceedings of the 52nd Annual Meeting of MSA, New Orleans* (eds GW Bailey, AJ Barrat-Reed). San Francisco Press, San Francisco, pp. 132–133.

Harris JR. (1996) Immunonegative staining: epitope localization on macromolecules. *Methods*, in press.

Harris JR, Engelhardt H, Volker S, Holzenburg A. (1993a) Electron microscopy of human erythrocyte catalase: new two-dimensional crystal forms. *J. Struct. Biol.* **111**, 22–33.

Harris JR, Gebauer W, Markl J. (1993b) Immunoelectron microscopy of keyhole limpet hemocyanin: a parallel subunit model. *J. Struct. Biol.* **111**, 96–113.

Hayat MA. (1989–1991) *Colloidal Gold: Principles, Methods and Applications*, Vols 1–3. Academic Press, New York.

Hyatt A. (1991) Immunonegative staining. In *Electron Microscopy in Biology: a Practical Approach* (ed. JR Harris). IRL Press, Oxford, pp. 59–81.

Kleinschmidt AK. (1968) Monolayer techniques in electron microscopy of nucleic acid molecules. In *Methods in Enzymology* (eds L Gross, K Moldave) Vol. 12B. Academic Press, New York, pp. 361–379.

Kleinschmidt JA, Hugle B, Gründ C, Franke WW. (1983) The 22S cylinder particles of *Xenopus laevis*. I. Biochemical and electron microscopical characterization. *Eur. J. Cell Biol.* **32**, 143–156.

Koeck PJB, Leonard KR. (1996) Improved immuno double labelling for cell and structural biology. *Micron*, in press.

Kopp F, Dahlman B, Hendil KB. (1993) Evidence indicating that the human proteasome is a complex dimer. *J. Mol. Biol.* **229**, 14–19.

Kopp F, Kristensen P, Hendil KB, Johnsen A, Sobek A, Dahlman B. (1995) The human proteasome subunit HsN3 is located in the inner rings of the complex dimer. *J. Mol. Biol.* **248**, 264–272.

Lambert O, Boisset N, Pochon F, Delain E, Lamy JN. (1994) An approach to the intramolecular localization of the thiol ester bonds in the internal cavity of human α_2-macroglobulin based on correspondence analysis. *J. Struct. Biol.* **112**, 148–159.

Lamy J, Billiald P, Taveau J-C, Boisset N, Motta G, Lamy J. (1990) Topological mapping of 13 epitopes on a subunit of *Androctonus australis* hemocyanin. *J. Struct. Biol.* **103**, 64–74.

Lünsdorf H, Tiedge H. (1992) Immunoelectron microscopy of enzymes, multienzyme complexes and selected other oligomeric proteins. *Electr. Microsc. Rev.* **5**, 105–127.

Martin J, Goldie KN, Engel A, Hartl FU. (1994) Topology of the morphological domains of the chaperonin GroEL visualized by immuno-electron microscopy. *Biol. Chem. Hoppe Seyler* **375**, 635–639.

Nermut MV. (1991) Unorthodox methods of negative staining. *Micron Microsc. Acta* **22**, 327–339.

Nermut MV, Frank H. (1971) Fine structure of influenza virus A2 (Singapore) as revealed by negative staining, freeze-drying and freeze-etching. *J. Gen. Virol.* **10**, 37–51.

Nermut MV, Perkins WJ. (1979) Consideration of the three dimensional structure of the adenovirus hexon from electron microscopy and computer modelling. *Micron* **10**, 247–266.

Oliver C. (1994) Conjugation of colloidal gold to proteins. In *Methods in Molecular Biology* (ed. C Javoic) Vol. 34. Humana Press, Totowa, NJ, pp. 303–307.

Roberts IM. (1986) Immunoelectron microscopy of extracts of virus-infected plants. In *Electron Microscopy of Proteins* (eds JR Harris, RW Horne) Vol. 5. Academic Press, London, pp. 293–357.

Schwartz H, Riede L, Sonntag I, Henning U. (1983) Degrees of relatedness of T-even type *E. coli* phages using different or the same receptors and topology of serological cross-reacting sites. *EMBO J.* **2**, 375–380.

Tanaka K, Ichihara A. (1990) Proteasomes (multicatalytic proteinase complexes) in eukaryotic cells. *Cell Struct. Funct.* **15**, 127–132.

Watts NRM, Hainfeld J, Coombs DH. (1990) Localization of the proteins gp7, gp8 and gp10 in the bacterophage T4 baseplate with colloidal gold: F(ab)$_2$ and undecagold: Fab' conjugates. *J. Mol. Biol.* **216**, 315–325.

Woodcock C, Baumeister W. (1990) Different representation of a protein structure obtained with different negative stains. *Eur. J. Biochem.* **51**, 45–52.

Yang Y-S, Datta, A, Hainfeld JF, Furuya FR, Wall JS, Frey PA. (1994) Mapping the lipoyl groups of the pyruvate dehydrogenase complex by use of gold cluster-labels and scanning transmission electron microscopy. *Biochemistry* **33**, 9428–9437.

5 Negative Staining: Selected Applications

A number of negative staining applications will be given below, taken from my own studies and from the published works of others. These represent a personal selection and no attempt has been made to present a comprehensive survey of the extremely widespread biological and biomedical applications of negative staining. Those requiring further detailed information can readily access the extensive literature via a computer topic or author search. It will be found that negative staining continues to be widely used, alone and increasingly as a preliminary or in parallel with cryoelectron microscopical studies of unstained frozen-hydrated specimens (Norcum *et al.*, 1994).

Images with and without computer processing will be presented here (see also Chapters 9 and 10). It is often apparent from the direct visual examination of TEM images that negative staining provides clear structural information and that an immediate and usually reliable interpretation can be made. When doing this it has to be accepted that the image from negative staining will often be at best in the order of 20 Å. The inclusion of computer digital image processing, either via single particle or crystallographic averaging, provides the possibility to retrieve the maximal available resolution within the images; this may result in 2-D and 3-D image reconstructions that approach a resolution of 15 Å from material studied at low temperatures in the presence of protective carbohydrate (glucose or trehalose) and negative stain. It has even been claimed that negative staining may achieve a resolution better than 10 Å in the near future (Marin van Heel and Prakash Dube, personal communication). Whether this would take negative staining beyond the widely accepted limit of providing information only on the surface contour/shell of a biological structure is not entirely clear, but it does indicate that a more general improvement might be achievable to this intermediate resolution level, without having to accommodate the problem of low signal-to-noise and defocus phase contrast that are always implicit in the study of unstained frozen-hydrated specimens.

Examples will be given from several negative staining studies on membranes, lipid-membrane systems, filamentous and tubular structures, viruses and macromolecules. Within the text, only brief comment will usually be made on biological aspects of the structures presented. Where possible references will be included to enable the reader to seek additional information of a structural and functional nature, or further details of the specimen preparations.

5.1   Membranes and lipid-membrane systems

The blood platelet, derived by megakaryocyte fragmentation *in vivo*, is one of the few intact cellular structures that is thin enough to be successfully imaged by negative staining. *Figure 5.1a* shows a single shape-transformed human blood platelet, negatively stained with 2% ammonium molybdate. Excessive electron density is present in places, but the more flattened periphery and extending filopodia, each containing a bundle of actin filaments, show clearly. In *Figure 5.1b*, at slightly higher magnification, the membrane surface of the extending processes indicates the presence of the particulate glycocalyx. Extraction with neutral surfactant enables the cytoskeleton of the blood platelet and thinly spread cultured cells to be revealed by negative staining (Hyatt, 1991; Small *et al.*, 1994; White, 1991).

The haemoglobin-free mammalian erythrocyte 'ghost' (Harris and Agutter, 1970) when collapsed and embedded in negative stain on a carbon support film presents only a double-membrane thickness. Such plasma membrane material is ideal for the investigation of lectin binding and the action of membrane-active toxins (Bhakdi *et al.*, 1985; Tranum-Jensen, 1988). *Figure 5.2* shows a single human erythrocyte ghost following treatment with the toxin streptolysin-O (SLO), monomer molecular mass 60 kDa, which oligomerizes to create approximately 24 nm ring-like lesions in the membrane bilayer. This figure therefore represents a *holey ghost*; with close inspection the curved edge of the membrane will be seen to lack the usual continuous sharply defined fold and over the whole of the membrane small electron dense cavities are present. At higher magnifications, preferably with a single layer of torn membrane, somewhat greater detail of the toxin lesions and their variability can be detected (*Figure 5.3*). SLO has long been believed to interact only with cellular membranes that contain cholesterol, a fact confirmed by the ability of this bacterial toxin to interact with liposomes composed of phosphatidylcholine–cholesterol, but not with pure phosphatidylcholine liposomes. Even stronger direct evidence comes from the fact that SLO will interact with pure cholesterol in the form of an aqueous suspension of microcrystals or as cholesterol bilayers and multilayers adsorbed from mica on to a carbon support film (Harris, 1988; see also Section 3.5). Cholesterol binding is rapidly followed by the oligomerization process, apparently triggered specifically by cholesterol (Duncan and Schlegel, 1975), leading to growth of the toxin arcs and production of the ring-like lesions. Negatively stained cholesterol microcrystals are shown in *Figure 5.4*. After interaction with SLO, the surface of the cholesterol becomes coated with lesions, which can break away from the

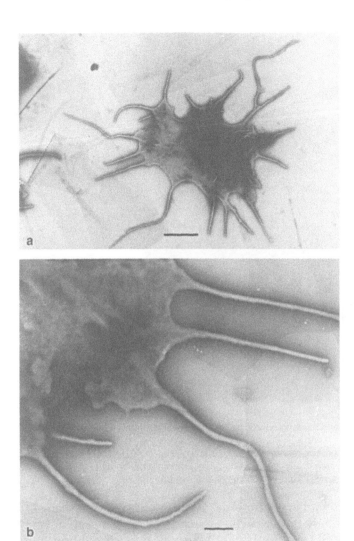

Figure 5.1: Negatively stained human blood platelets, after shape transformation.

Human blood platelets negatively stained with 2% ammonium molybdate (pH 7.0) by the droplet technique. Regions of excessive thickness/density are present, but the surface of the extending finger-like processes, each with a central actin bundle, shows the presence of the particulate surface glycocalyx (b). The scale bars indicate 1 µm (a) and 200 nm (b).

surface and also form large lace-like networks composed of inter-linked ring-like lesions (*Figures 5.5.* and *5.6.*). Whether or not the individual lesions and the toxin networks contain some hydrophobically bound cholesterol or are composed of oligomerized toxin alone has yet to be determined; the former seems to be a distinct possibility. With the massive expansion of molecular biology and molecular genetics, cloned mutant proteins, including toxins, are becoming widely available for comparison structurally

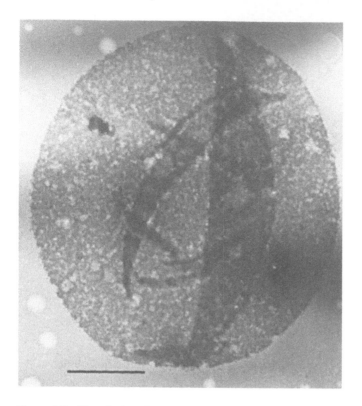

Figure 5.2: Negatively stained human erythrocyte ghost, following treatment with the bacterial toxin streptolysin O.

The originally smooth unbroken membrane surface now appears ragged and is pitted with many small circular lesions. Negatively stained with 5% ammonium molybdate containing 1% trehalose (pH 7.0) by the droplet procedure. The scale bar indicates 2 μm.

and functionally with wild-type proteins. Here too, in many instances, electron microscopic study of negatively stained specimens may have a contribution to make. SLO mutants can be produced with reduced and even minimal capacity to produce cell lysis, but still with the potential to bind to natural and artificial membranes containing cholesterol or indeed pure cholesterol, but not phospholipids alone. In the latter instance, the bound mutant SLO (asp402→cys) can be detected as a surface coating on cholesterol microcrystals, which may create an orderly binding pattern or single layer toxin 2-D array (*Figure 5.7a*) with no ring-like lesions. If the bound SLO has itself been biotinylated prior to interaction with the cholesterol, subsequent labelling with avidin- or streptavidin-conjugated colloidal gold particles can confirm the presence of the bound SLO mutant (*Figure 5.7b*).

Erythrocyte ghosts and the surface/plasma membrane of other cells will also bind or interact with numerous other toxins, plant, and animal lectins, serum complement and lytic agents such as the saponins. These agents or their structural reaction product can be detected as individual molecules or oligomeric complexes, which are often much smaller than the SLO lesions. The *Staphylococcus aureus* α-toxin is an example (*Figure 5.8*). This toxin (monomer 34 kDa) oligomerizes on the membrane to form a heptameric group with affinity for the membrane bilayer through which it generates a channel, thereby producing cytolysis; in this case membrane phospholipid rather than cholesterol is thought to be the site of interaction. *S. aureus* α-toxin will interact with phosphatidylcholine liposomes (*Figure 5.9*) upon which the bound heptameric complexes can be seen. Control liposomes are much smoother, more regular concentric bilayer structures, clearly lacking any indication of bound particulate material (*Figure 5.10*). An elegant study of the related heptameric channel-forming toxin aerolysin (from *Aeromonas hydrophila*) by Wilmsen *et al.* (1992) demonstrates further the possibilities for the creation of 2-D crystals of this toxin and the computer processing of images from negatively stained specimens (*Figures 5.11* and *5.12*). This work has been supported by the higher resolution X-ray structure (Parker *et al.*, 1994), whereas the electron microscopical studies on the *S. aureus* α-toxin (Hebert *et al.*, 1992; Ward and Leonard, 1992) indicated six-fold rather than the now accepted

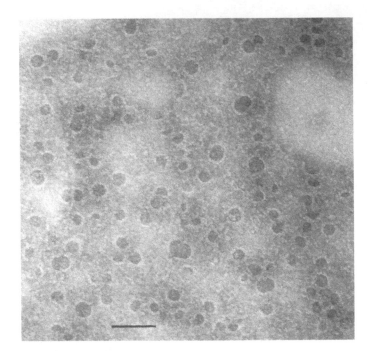

Figure 5.3: A higher magnification of an erythrocyte ghost.
 A higher magnification from part of the surface of a broken membrane of an erythrocyte ghost, following treatment with the toxin streptolysin O. The arc-like and circular toxin lesions, which create stain-filled holes, are clearly defined. Negatively stained with 5% ammonium molybdate containing 1% trehalose (pH 7.0). The scale bar indicates 100 nm.

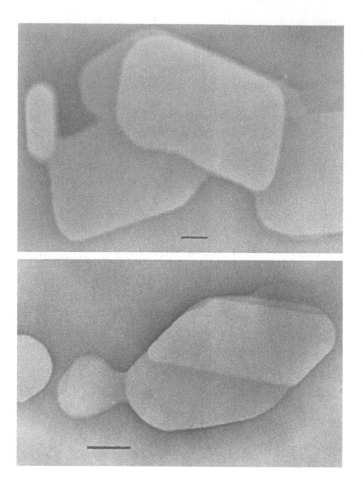

Figure 5.4:   Negatively stained cholesterol microcrystal clusters.

Cholesterol microcrystal clusters negatively stained with 5% ammonium molybdate containing 1% trehalose (pH 7.0). (Cf. Harris, 1988, and *Figure 8.1* showing unstained, frozen-hydrated cholesterol.) The scale bars indicate 100 nm.

seven-fold symmetry for the oligomer (Gouaux *et al.*, 1994).

Other isolated cellular membranes such as the plasma membrane and gap junctions from tissue cells (Benedetti and Emellot, 1968), rough and smooth endoplasmic reticulum, Golgi membranes and annulate lamellae have all received attention from electron microscopists and an early important contribution was made by negative staining to the understanding of the two membranes and compartmentation within the mitochondrion. The inner mitochondrial membrane is usually a tightly packed tubular or lamellar structure (the *cristae*), characteristically coated with small 'knobs' known to be the F_1–ATPase complex (*Figure 5.13*). There is now a considerable body of structural data on the isolated F_1–ATPase complex from negative staining (Boekema *et al.*, 1986; Fujiyama *et al.*, 1990) and the

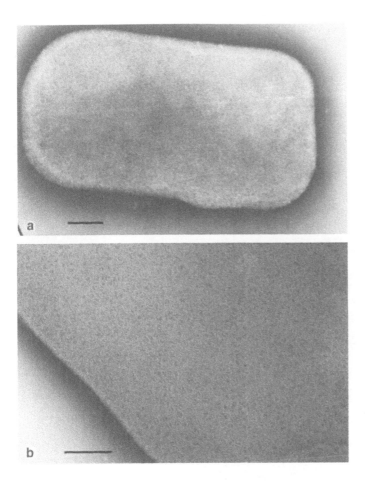

Figure 5.5: Negatively stained cholesterol microcrystals after interaction with the toxin streptolysin O (a and b).
The cholesterol surface is coated with circular lesions, which in (b) have formed a complex overlapping network. Negatively stained with 5% ammonium molybdate containing 1% trehalose (pH 7.0). The scale bars indicate 200 nm.

high resolution X-ray structure has recently been determined (Abrahams *et al.*, 1994). Studies on the voltage-dependent anion-selective channel from the mitochondrial outer membrane have been performed (Guo and Mannella, 1992; Guo *et al.*, 1995) using 2-D crystals embedded in aurothioglucose and vitreous ice. Additionally, Hofhaus *et al.* (1991) have used negative staining with uranyl acetate to analyse the peripheral and membrane parts of the mitochondrial NADH dehydrogenase.

The cell nucleus is too thick to be usefully studied by negative staining, but the surface double-membrane envelope of the nucleus and its associated nuclear pore complexes have been successfully investigated by negative staining following isolation by chromatin decondensation and deoxy-

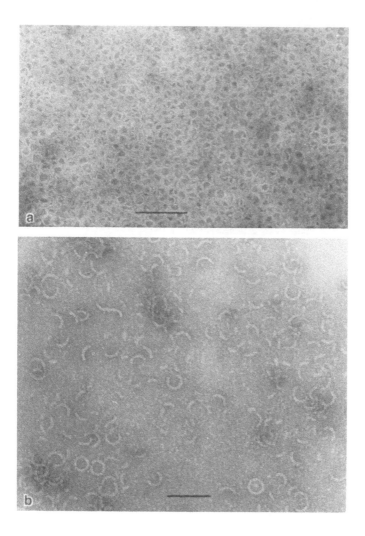

Figure 5.6:   Negatively stained streptolysin O lesions on a cholesterol crystal.

Part of a streptolysin O lesion-network, that has separated from a cholesterol microcrystal (a) and a region of well dispersed arc-like and circular streptolysin O lesions, which can be detected alongside cholesterol microcystals (b) (cf. *Figure 5.3*). Negatively stained with 5% ammonium molybdate containing 1% trehalose. The scale bars indicate 200 nm (a) and 100 nm (b).

ribonuclease digestion (*Figure 5.14a*). The appearance of the nuclear pore complexes varies somewhat in different stains (Harris, 1978; Harris and Marshall, 1981); it is now generally thought that ammonium molybdate provides the most reliable images. The eight-fold rotational symmetry of the nuclear pore complex (*Figure 5.14b*), suggested by early thin sectioning studies as well as negative staining, has been confirmed and extended by recent studies on unstained frozen-hydrated nuclear envelope (see Chap-

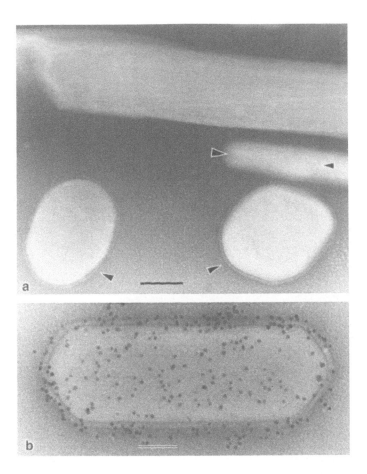

Figure 5.7: Negatively stained cholesterol microcrystals after interaction with a streptolysin O mutant.

Cholesterol microcrystals following interaction with a streptolysin O mutant (asp402→cys), defective cytolytically (lesion-forming ability). The cholesterol surface can be seen to be coated with material (arrows, a), which appears to consist of a monomolecular layer containing an ordered linear array of toxin. This is shown more clearly (b) at higher magnification, with biotinylated toxin mutant, subsequently labelled with 5 nm streptavidin–gold particles. Negatively stained with 5% ammonium molybdate containing 1% trehalose (pH 7.0). The scale bars indicate 100 nm (a) and 50 nm (b).

ter 8). Procedures for the release of the nuclear pore complex from the inner and outer nuclear membrane have met with major technical difficulties over a lengthy period. Neutral surfactant extractions followed by controlled disruption by ultrasonication (Marshall and Harris, 1979) can lead to the release of the ring-like nuclear pore complexes (*Figure 5.15*). Unfortunately, these isolated nuclear pore complexes clearly exhibit varying amounts of structural damage, but improved biochemical isolations

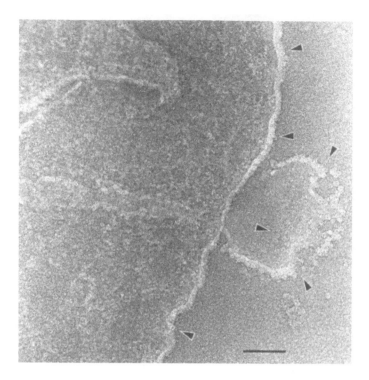

Figure 5.8:   Negatively stained rabbit erythrocyte ghost after interaction with *Staphylococcus aureus* α-toxin.

Part of the surface of a rabbit erythrocyte ghost following interaction with *S. aureus* α-toxin. Note the presence of small particles (heptamers) coating the membrane surface (arrows). These are more clearly visible on the thinner vesicular projection at the RHS. Negatively stained with 5% ammonium molybdate containing 1% trehalose (pH 7.0). The scale bar indicates 100 nm.

should eventually be possible. Much investigation at the cell biological, biochemical and structural level is in progress on the nuclear pore complex during nuclear envelope formation *in vitro*, and negative staining continues to make a useful contribution, including immunonegative staining (Guan *et al.*, 1995).

In vitro reformation of membrane systems and 2-D membrane crystals, from extracted and purified membrane proteins and from proteins cloned in bacteria, has been an area to which negative staining has been extensively applied. Reformation of the gap junction has been experimentally achieved (Ghoshroy *et al.*, 1995; Kistler *et al.*, 1993, 1994; Konig and Zampighi, 1995; Sosinsky, 1995; Stauffer *et al.*, 1991). The progressive formation of mini-gap junctions and 2-D crystals from purified hemi-channels can be monitored readily by negative staining (see *Figure 5.16*), an approach utilized successfully by Andreas Engel and his colleagues for this and other membrane systems. The understanding of the protein or-

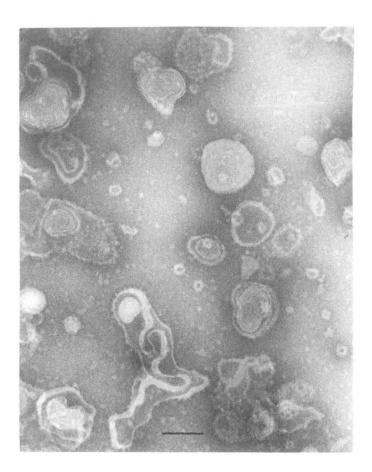

Figure 5.9: Negatively stained liposomes after interaction with *S. aureus* α-toxin.
    Liposomes containing phosphatidylcholine and cholesterol, following interaction with *S. aureus* α-toxin. The surface of the outermost lipid bilayers is coated with the heptameric oligomers (previously thought to be hexamers; Bhakdi *et al.*, 1993; Ward and Leonard, 1992). With this toxin, it is thought that the phospholipid is the site of interaction, rather than cholesterol. Negatively stained with 5% ammonium molybdate containing 1% trehalose (pH 7.0). The scale bar indicates 100 nm.

ganization within the specialized luminal plasma membrane of the mammalian urinary bladder (Knutton, 1982) is also an area where negative staining continues to make a major contribution (Walz *et al.*, 1995).

The study of the chloroplast, bacterial photosystems and membrane enzymes is another area to which negative staining of 2-D protein–lipid crystals has made an important contribution (Boekema *et al.*, 1988, 1989, 1994; Welte *et al.*, 1995). Two-dimensional and 3-D image processing has been performed from 2-D crystals of the photosystem II from spinach chloroplasts (*Figure 5.17*) by Andreas Holzenburg and his colleagues (Ford *et al.*, 1995; Holzenburg *et al.*, 1994). Although this work may to some

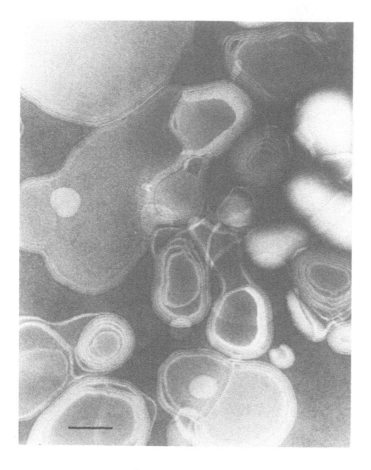

Figure 5.10: Negatively stained control liposomes.
Control phosphatidylcholine–cholesterol multi-bilayer liposomes, negatively stained with 5% ammonium molybdate containing 1% trehalose (pH 7.0). The scale bar indicates 100 nm.

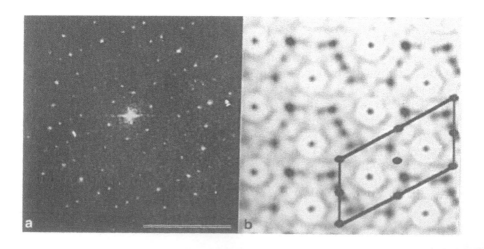

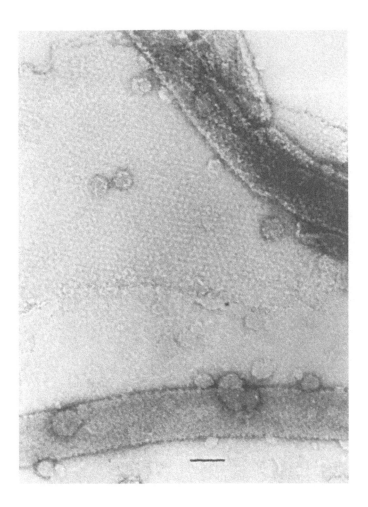

Figure 5.11: Negatively stained aerolysin following interaction with *E. coli* phosphatidylethanol-amine.

A 2-D crystal and helical tubes of aerolysin (from *Aeromonas hydrophilia*) following interaction with *E. coli* phosphatidylethanolamine (see Wilmsen *et al.*, 1992, Figure 4). Negatively stained with 1% sodium phosphotungstic acid. The scale bar indicates 100 nm. Previously unpublished electron micrograph, courtesy of Kevin Leonard.

◄ **Figure 5.12:** Fourier transform of aerolysin.

(a) A computed Fourier transform from a negatively stained 2-D crystalline sheet of aerolysin. The scale bar indicates a reciprocal distance of 1/25 Å. (b) A noise-filtered, average image calculated from the peak Fourier coefficients. The unit cell dimensions are 133 Å × 264 Å, cell angle 118°; two-fold axes of the p2 unit cell are marked. The asymmetric unit is a heptameric 'wheel-like' structure with a strongly staining central ring surrounding a darker stain-filled depression or channel approximately 17 Å in diameter. Figure courtesy of Kevin Leonard. Modified from Wilmsen *et al.* (1992) The aerolysin membrane channel is formed by heptamerization of the monomer. *EMBO J.* 11, 2457–2463, by permission of Oxford University Press.

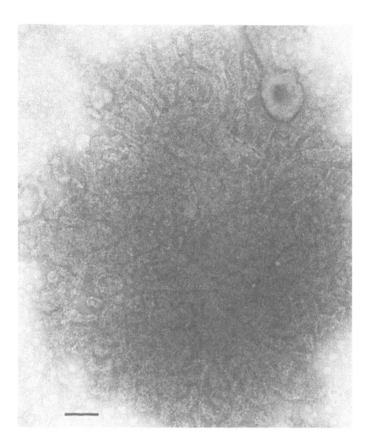

Figure 5.13:   Negatively stained rat liver inner mitochondrial membranes.

A cluster of rat liver inner mitochondrial membranes (cristae), following release from the outer mitochondrial membrane, showing the surface coating of F_1–ATPase. Negatively stained with 2% sodium phosphotungstate (pH 7.0). The scale bar indicates 200 nm.

extent be overshadowed by higher resolution studies from unstained material (Kühlbrandt *et al.*, 1994) these may eventually be surpassed by X-ray diffraction studies (McDermott *et al.*, 1995). Although most studies on ordered bacterial membranes have centred around the purple membrane of *Halobacterium halobium* and the high resolution studies in glucose (Henderson *et al.*, 1990), it should perhaps be recalled that negative staining can reveal the 2-D lattice of this readily available natural membrane crystal (Nermut, 1991), albeit at lower resolution.

The two-dimensionally ordered bacterial glycoprotein surface layer (S-layer), as distinct from lipid-rich membrane structures, is a further area to which negative staining has made a major contribution (Baumeister and Engelhardt, 1987; Beveridge and Graham, 1991; Durr *et al.*, 1991;

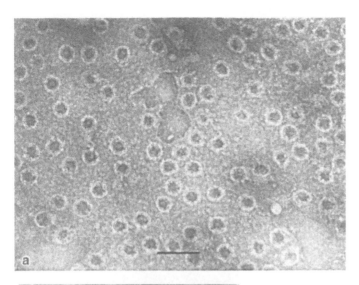

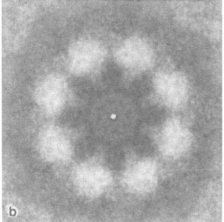

Figure 5.14: Negatively stained rat liver nuclear envelope.

A fragment of rat liver nuclear envelope negatively stained with 2% ammonium molybdate (pH 7.0). The nuclear pore complexes are clearly revealed, as is the particulate coating of ribosomes on the surrounding outer nuclear membrane (a). Rotational averaging indicates the eight-fold symmetry of the nuclear pore complex (b). (Cf. *Figures 8.3* and *8.4* and see Harris and Marshall, 1981.) The scale bar (a) indicates 200 nm.

Pum *et al.*, 1993; Sara and Slytr, 1996). The subtly varying forms of S-layer can be classified and the structure of the fundamental units assessed by 2-D and 3-D image processing, as exemplified by *Figure 5.18* from the bacterium *Thermus thermophilus* (Castón *et al.*, 1993).

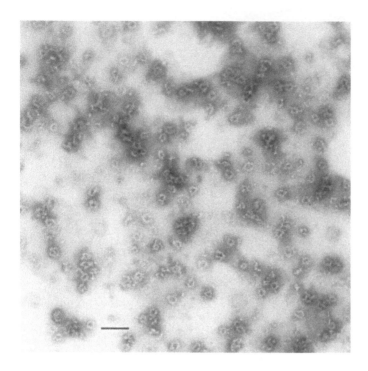

Figure 5.15: Negatively stained nuclear pore complexes.
Nuclear pore complexes isolated from nuclear envelope by neutral surfactant extraction and ultrasonication, followed by sucrose gradient centrifugation. Negatively stained with 2% ammonium molybdate (pH 7.0). See Marshall and Harris (1979). The scale bar indicates 300 nm.

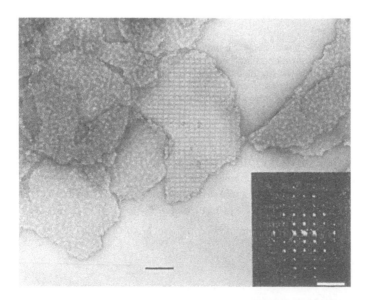

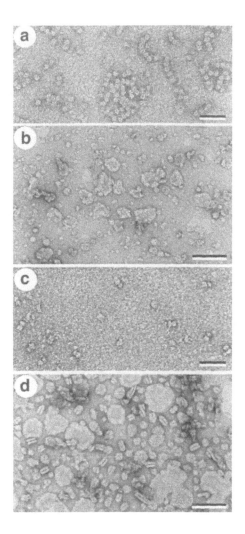

Figure 5.16: Negatively stained gap junctions.

Crude assembly of gap junction from hemichannels (single connexons), negatively stained with 2% uranyl acetate. (a) 8-POE-solubilized hemichannels, which appear as doughnut-shaped particles. (b) Small membrane sheets reconstituted from 8-POE-solubilized lens fibre membrane proteins and lipids. Note the absence of double-layered structures. (c) Pairing of hemichannels by exchange of 8-POE with 10-MALT produces connexon pairs appearing as dumbbell-shaped particles. (d) Reconstitution of paired connexons with fibre membrane lipids yielding abundant double-layered mini-gap junctions. The scale bars indicate 50 nm (a and c) and 100 nm (b and d). Micrographs courtesy of Andreas Engel. Reproduced from Kistler *et al.* (1994) Reconstitution of native-type non-crystalline lens fiber gap junctions from isolated hemichannels. *J. Cell Biol.* 126, 1047–1058, by copyright permission of The Rockefeller University Press.

◀ **Figure 5.17:** Two-dimensional crystal of thylakoid photosystem II.

A reconstituted Tris buffer washed 2-D crystal of thylakoid photosystem II, negatively stained with 4% uranyl acetate, together with the power spectrum (inset). The scale bars indicate 0.2 nm^{-1}(inset) and 100 nm (micrograph). Micrograph courtesy of Andreas Holzenburg. Reprinted from Ford *et al.* (1995) Photosystem II 3-D structure and the role of the extrinsic subunits in photosynthetic oxygen evolution. *Micron* 26, 133–140, with permission from Elsevier Science Ltd.

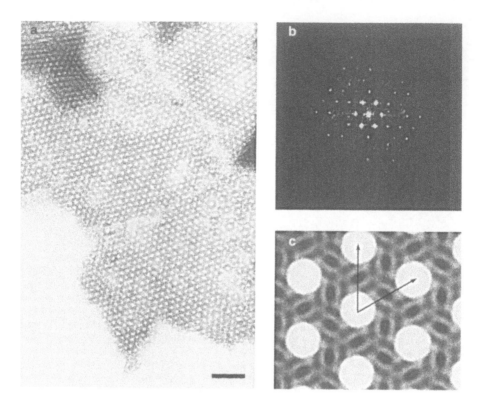

Figure 5.18: 2-D analysis of a S1 2-D crystal of *Thermus thermophilus* HB8 S-layer protein.
(a) 2-D crystal negatively stained with 2% uranyl acetate. (b) The optical transform of the 2-D crystal, with diffraction spots extending to 0.33 nm⁻¹. (c) Two-dimensional averaged image of 10 Fourier filtered images of S1 crystals, representing 25 different areas each containing about 177 unit cells. The unit cell vectors, indicated by arrows, are a = b = 19 ± 0.2 nm. Scale bar (a) indicates 100 nm. Courtesy of José Carrascosa. Reproduced from Castón *et al.* (1993) S-layer protein from *Thermus thermophilus* HB8 assembles into porin-like structures. *Molec. Microbiol.* 9, 65–75, with permission from Blackwell Science Ltd.

5.2 Filamentous and tubular structures

Within the broad field of studies on fibrous and tubular assemblies negative staining has made a principal contribution, particularly for the understanding of collagen, intermediate filament, actin, myosin and associated muscle proteins (Bremner *et al.*, 1994; Egelman and Orlova, 1995; Lehman *et al.*, 1994) cytoplasmic and flagellar microtubules (Fok *et al.*, 1994), and bacterial pili (Bullitt and Makoowlski, 1995). In *Figure 5.19* part of a keyhole limpet (*Megathura crenulata*) haemocyte cytoplasmic microtubule bundle is shown, with attached microtubule associated proteins (MAPs) and also part of a sperm tail displaying the microtubule dou-

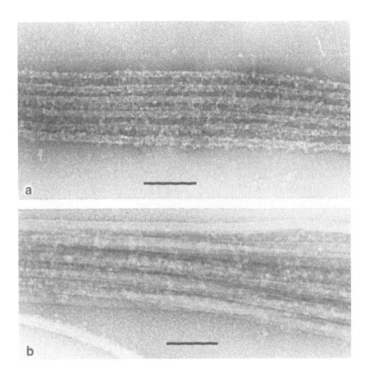

Figure 5.19: Negatively stained microtubules.
(a) A microtubule bundle from *Megathura crenulata* haemocytes, showing pronounced coating with microtubule associated proteins (MAPs). (b) Part of a disrupted sperm tail, showing the parallel microtubule doublets. Both negatively stained with 5% ammonium molybdate containing 1% trehalose (pH 7.0). The scale bars indicate 200 nm.

blets splaying out, in this instance from the main body of an intact 9+2 structure (Harris *et al.*, 1996). Microtubules reconstituted from pig brain tubulin are shown in *Figure 5.20*, negatively stained with uranyl acetate–trehalose and ammonium molybdate–trehalose. The experimental decoration of microtubules with MAPs and the kinesin motor protein have been successfully investigated by negative staining (Amos, 1982; Hirose *et al.*, 1995; Hoenger *et al.*, 1995), as discussed further in Chapter 10. In the absence of trehalose considerable microtubule flattening occurs, which is largely prevented by the thicker embedding layer when 1% trehalose is included. When embedded in ammonium molybdate–trehalose these microtubules are unstable and show marked indications of partial dissociation (*Figure 5.20b*). A similar situation also occurs with the helical cytoplasmic filament from *M. crenulata* haemocytes (Harris and Markl, 1992), as shown in *Figure 5.21*. Although the first reaction to such stain-induced disruption might be that this is totally unsatisfactory, progressive disso-

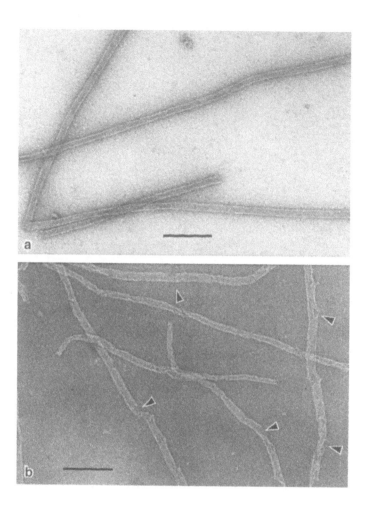

Figure 5.20:   Stain-induced disruption in pig brain microtubules.
Reconstituted pig brain microtubules negatively stained with 4% uranyl acetate containing 1% trehalose (a) and 5% ammonium molybdate containing 1% trehalose (b). Note the excellent preservation of microtubule structure in (a), whereas pronounced *stain-induced* disruption is apparent in (b), with breakage of the protofilaments (arrows) (cf. *Figure 8.5*). The scale bars indicate 200 nm. Microtubules courtesy of Max Gerber.

ciation or breakdown of a filamentous or tubular structure may provide subtle insight into the higher oligomeric/polymeric organization (see also Harris *et al.*, 1995b).

Collagen has a well characterized periodic paracrystalline structure, due to a precise parallel alignment and linear overlap of the elongated collagen monomer, and has been studied intensively by electron microscopists using negative staining (Chapman *et al.*, 1990). The collagen fibre periodicity is particularly well shown by negative staining with ammonium molybdate trehalose (*Figure 5.22*), which in this instance also indicates the progressive tapering of the fibres to extremely thin fibril termini. Paracrystalline arrays have been produced from many elongated and filamentous proteins, several of which have been studied by negative staining. One example is that of the nuclear lamins (Heitlinger *et al.*, 1991), as shown in *Figure 5.23*. Again, the production of mutant proteins can usefully provide insight into the individual amino acids and sequences of importance for the production of filamentous interactions between protein monomers. It should perhaps be borne in mind that tubular and filamentous arrays of proteins can be formed during the crystallization process (e.g. sickle cell haemoglobin and human erythrocyte catalase; Kiselev *et al.*, 1967; McDade and Josephs, 1993). Recently it has been found that the elongated multi-domain peptide of keyhole limpet haemo-

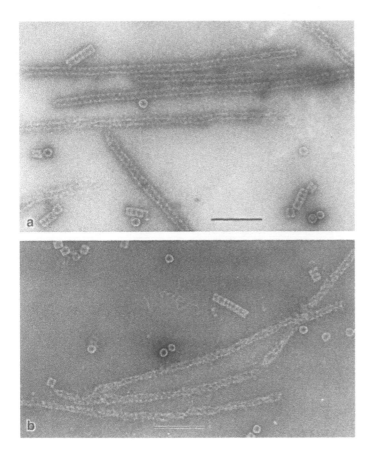

Figure 5.21: Helical filaments from *M. crenulata* haemolymph.

Helical filaments from *M. crenulata* haemolymph (Harris and Markl, 1992), negatively stained with 4% uranyl acetate containing 1% trehalose (a) and 5% ammonium molybdate containing 1% trehalose (pH 7.0). As with the microtubules shown in *Figure 5.20*, preservation of the helical filaments is superior in uranyl acetate (a); some disruption is indicated in the presence of ammonium molybdate (b). The scale bars indicate 200 nm.

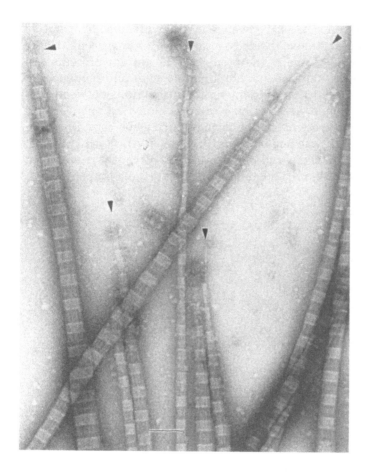

Figure 5.22: Negatively stained collagen fibres from *M. crenulata*.

Gradually tapering collagen fibres from *M. crenulata*, negatively stained with 5% ammonium molybdate containing 1% trehalose. Note the finely pointed ends of the fibres, which contain very few individual linearly overlapping collagen molecules (arrows). The scale bar indicates 100 nm. Modified from Harris *et al.* (1996).

cyanin type 1 (KLH1) can reassociate/polymerize into well defined twisted ribbons/helical tubules, and that these can aggregate to form paracrystalline bundles (*Figure 5.24*).

5.3   Viral structure

The detection of viruses and the investigation of their structure was one of the first scientific areas to which negative staining made a major impact (see Horne and Wildy, 1979). These structural data contributed significantly to the extensive classification of animal, plant and bacterial viruses that exists today. In the sphere of viral diagnostics the overall contribution of the transmission electron microscope has greatly diminished in recent years, except perhaps for the detection of reoviruses (Madely and Carter, 1986) and viral vaccine quality assessment. At the more fundamental level TEM has long been important (Wrigley, 1979; Wrigley *et al.*, 1986) and indeed continues to be so due to both negative staining and cryoelectron microscopy of unstained viruses in vitreous ice (see Chapters 8 and 10). The increasing availability of monoclonal antibodies to viral proteins now opens the door to extensive immunonegative staining stud-

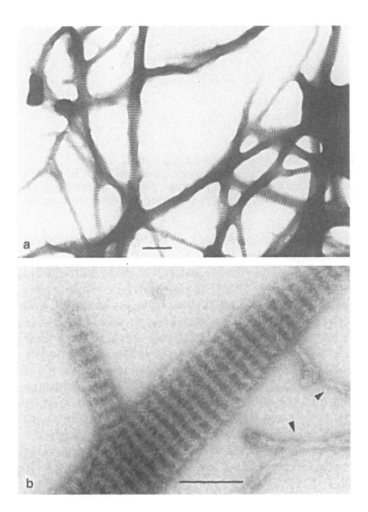

Figure 5.23: Negatively stained paracrystalline bundles of bacterially expressed lamin A.
Paracrystalline bundles of bacterially expressed lamin A, produced by the NS–CF technique (Harris, 1991), with 2% uranyl acetate as the negative stain. Note the approximately 25 nm periodicity (b) and the presence of small lamin A aggregates alongside the large paracrystalline bundles (arrows, b). The scale bars indicate 250 nm (a) and 100 nm (b).

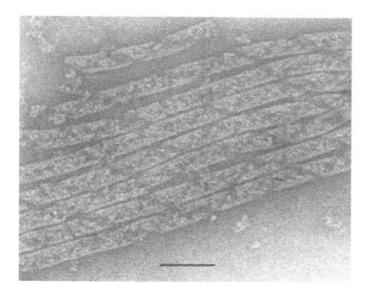

Figure 5.24:   Negatively stained parallel array of helical tubules of KLH1.
A parallel array of helical tubules produced by reassociation/polymerization of KLH1 subunits in the presence of a high calcium and magnesium concentration. Negatively stained with 5% ammonium molybdate containing 1% trehalose (pH 7.0) (Harris *et al.*, 1996). The scale bar indicates 100 nm.

ies as well as immuno-cryoelectron microscopy for epitope localization and subsequent structural interpretation.

A classical example, of a fowlpox virus, negatively stained with sodium phosphotungstate is given in *Figure 5.25*. The characteristic block-shaped structure of this virus, with undulating surface tubules, has been well documented. Generally, the TEM images of viruses embedded in negative stain are thought to be valid, if low resolution, representations of their true shape and surface features. This claim has often been supported by metal shadowing studies and more recently by cryoelectron microscopy of unstained specimens. There are, however, examples where this is not the case, for example vaccinia virus (an orthopoxvirus), as documented by Dubochet *et al.* (1994). This could be due to a lack of negative stain penetration within the surface membrane or capsid, or actually inside the viral core; pH-induced viral shrinkage and distortion within the negative stain before stain drying and embedding could also be another reason for this discrepancy.

For the purpose of a more detailed yet brief comparison between negative staining and cryoelectron microscopy of unstained vitrified viruses, several TEM images of adenovirus will be presented here and also in Chapters 8 and 10. It is my intention to emphasize here the contribution of early negative staining data to the understanding of the structure of this group of viruses (see also Nermut, 1991, and Horne, 1986). Examples of

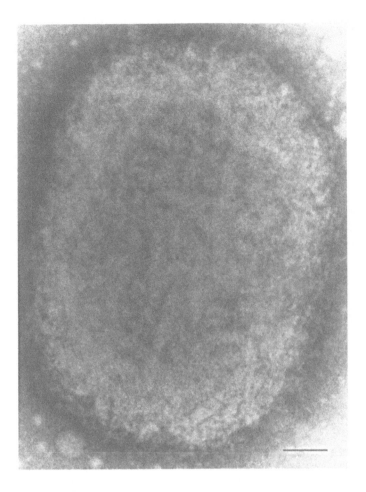

Figure 5.25: Negatively stained fowlpox virus.
An example of a single fowlpox virus, negatively stained with 2% sodium phosphotungstate (pH 7.0). Note the undulatory surface tubule-like feature. The scale bar indicates 50 nm.

human adenovirus type 5 are shown in *Figure 5.26*, negatively stained with sodium silicotungstate by the freeze-dry negative staining procedure (Chapter 4). Note the presence of some disrupted virions. At higher magnification detail of the capsid hexon arrangement can be seen within the uppermost viral facet, which correlates with the well established group of nine (GON) hexons that tend to split away when the viral capsid undergoes disruption (*Figure 5.27*). Of considerable significance is the fact that the handedness of the group of nine hexons can be defined by negative staining. In relatively shallow stain only the portion of the isolated hexon of GON closest to the carbon support film contributes to the image (Nermut and Perkins, 1979). Thus, in addition to the handedness it was possible to define a shape difference between the hexon base and the hexon surface region. Depending upon the charge of the carbon support film, the group

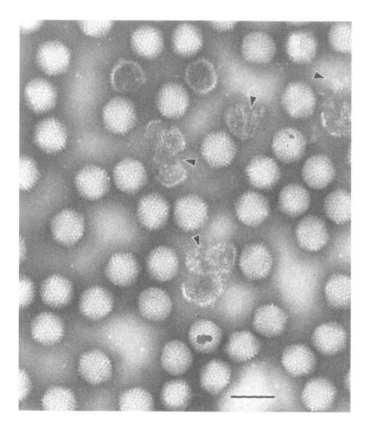

Figure 5.26: Negatively stained human adenovirus type 5.
Human adenovirus type 5, negatively stained with sodium silicotungstate by the freeze-dry negative staining procedure (Chapter 4). Note the presence of disrupted viral capsids and the release of groups of nine (GON) hexons (arrows). The scale bar indicates 50 nm. Micrograph courtesy of Milan Nermut (cf. *Figures 8.15* and *8.16*).

of nine hexons became selectively adsorbed with the surface aspect of the hexons (negatively charged) or the hexon base (hydrophobic) facing the carbon. The more circular hexon capsid base was then distinguished from the more triangular hexon surface, within the group of nine hexons (inset a, *Figure 5.27*). The left hand orientation represents the GON orientation as seen directly within the 12 hexons of each facet of the intact virion. Negative staining has also contributed to the study of the isolated hexon, the penton base and penton fibre of adenovirus (Albiges-Rizo *et al.*, 1991; Boudin *et al.*, 1979; Hess *et al.*, 1995; Misel, 1978; Nermut and Perkins, 1979; Petterson and Höglund, 1969) and provided information of significant value for the subsequent interpretation of unstained cryoimages and indeed X-ray data (Milan Nermut, personal communication).

The production of 2-D arrays of adenovirus by the negative staining–carbon film technique has proved to be a successful approach for the in-

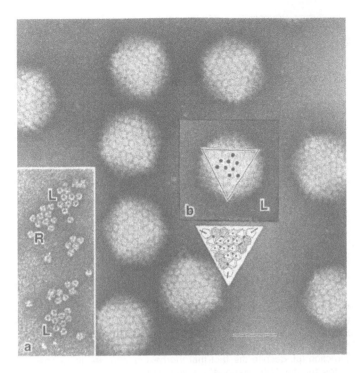

Figure 5.27: Negatively stained human adenovirus type 5.
A higher magnification micrograph showing human adenovirus type 5, negatively stained as in *Figure 5.26*. Inset (a) shows left- and right-handed GON hexons and inset (b) a viral facet, with the *left-handed* GON indicated (L) (cf. *Figure 8.16*). The scale bar indicates 50 nm. Micrograph courtesy of Milan Nermut. See text for further comment on structural detail.

vestigation of this virus (*Figure 5.28a*). Although a reasonable optical transform/fast Fourier transform/power spectrum can be obtained from such arrays (*Figure 5.28b*), it would appear that the 2-D viral array is not crystallographically highly ordered, since the reconstructed 2-D p622 image (*Figure 5.28c*) does not retain significant information at the hexon level in this instance (Horne, 1986; Horne *et al.*, 1975). It is also likely that this 2-D array of viruses may be incompletely embedded (partial depth coverage) in negative stain. This does *not necessarily imply* that under slightly different crystallization conditions perfect 2-D viral crystals could not be formed and that these could not be completely embedded in negative stain (Horne *et al.*, 1977; Steven *et al.*, 1978). Inclusion of PEG appears to be of vital importance (Wells *et al.*, 1981) together with the determination of viral concentration that creates a continuous monoviral layer upon the mica during drying and therefore appropriate conditions for 2-D crystal production, or alternatively at a higher viral concentration, for the production of 3-D crystals (*Figure 5.29*). The negative staining–carbon film technique also has considerable potential for the investigation of viral disruption and *in vitro* reformation (Harris and Horne, 1994). Indeed, it is of

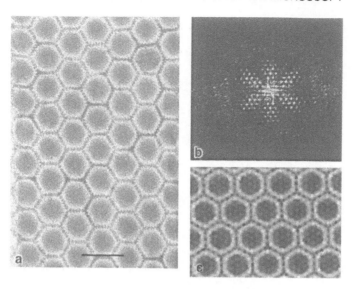

Figure 5.28: A 2-D array of human adenovirus type 5.

An example of a 2-D array of human adenovirus type 5, produced by the NS–CF procedure, with 0.5 % uranyl acetate as the negative stain (a), together with the power spectrum (b) and 2-D image average (b). Because of poor 2-D crystalline order, information is not retained at the hexon level in the reconstruction (c), despite the fact that the hexons can clearly be seen in (a).The scale bar (a) indicates 100 nm. Micrograph (a) courtesy of Bob Horne.

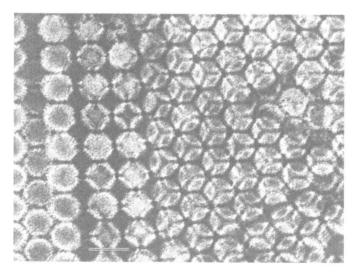

Figure 5.29: Part of a 3-D crystal of adenovirus type 5.

Part of a 3-D crystal of adenovirus type 5 produced by the NS–CF technique, from a high concentration of virus, with uranyl acetate as the negative stain. Varying viral packing and viral superimposition within the array creates image complexity, from which structural information cannot readily be retrieved by crystallographic image averaging. The scale bar indicates 100 nm. Micrograph courtesy of Bob Horne. Reprinted from Horne (1986) Electron microscopy of crystalline arrays of adenoviruses and their components. In *Electron Microscopy of Proteins* (eds JR Harris, RW Horne) Vol. 5. Academic Press, London, pp. 71–101, with permission from Academic Press Ltd.

considerably more than historical interest that Horne (1986) showed the formation of 2-D crystals from the GON adenovirus hexons, released following disruption of the virus during prolonged incubation in ammonium molybdate.

That 2-D, or indeed 3-D, crystallization on mica does in all probability occur at the fluid–air interface, is strongly suggested by the evidence from a thin sectioned carbon film with small 3-D crystals of tobacco mosaic virus (TMV) attached and surrounded by negative stain (*Figure 5.30*) (see Horne *et al.*, 1975, 1976). If this is the case, it has considerable implications for the understanding of the crystallization process and the further extension of this technique for the production of 2-D and/or small 3-D

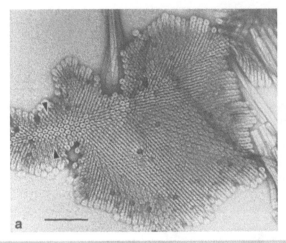

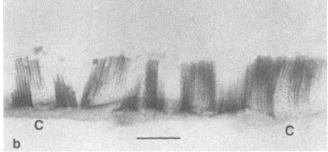

Figure 5.30:   A 3-D crystalline array of tobacco mosaic virus.
    (a) A 3-D crystalline array of tobacco mosaic virus (TMV), produced by the NS–CF technique, with 0.5% uranyl acetate as the negative stain. Note the limited region where the TMV rods are orientated vertically (between arrows), and the positions where individual TMV rods are absent from the crystal. Most TMV rods are obliquely orientated. (b) A thin section of such a 3-D crystalline array of TMV produced by the NS–CF procedure, following conventional resin embedding. The TMV rods are adsorbed to carbon film (C). Vertically orientated TMV rods define very well the average length of the virus (Ruiz *et al.*, 1994). The scale bars indicate 100 nm (a) and 200 nm (b). Both micrographs courtesy of Bob Horne.

crystals within ammonium molybdate alone, ammonium molybdate–PEG or PEG alone, either on mica, carbon films or across the holes of holey carbon support films (see Section 5.4; *Figures 5.47* and *5.48;* Chapter 9). A more recent application of the NS–CF technique to TMV by Ruiz *et al.* (1994) has demonstrated that when negatively stained with uranyl acetate, parallel, single and multilayer TMV arrays yielded an electron diffraction resolution of 3.8 Å.

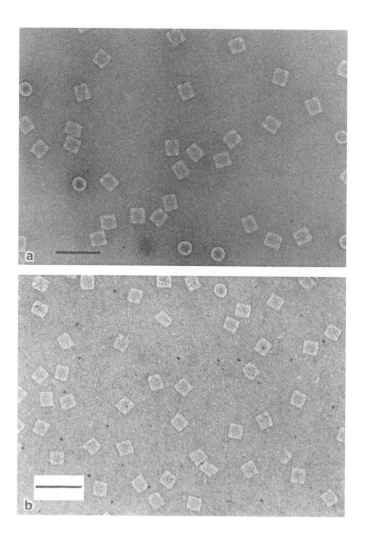

Figure 5.31: Negatively stained keyhole limpet haemocyanin type 1 (KLH1) didecamers. Purified KLH1 didecamers negatively stained with 5% ammonium molybdate containing 1% trehalose (pH 7.0). In (a) the specimen was studied at room temperature and (b) after specimen cooling to –175°C (see Harris *et al.*, 1995a). The scale bars indicate 100 nm.

5.4   Single molecules, 2-D and 3-D crystals

Negative staining has been successfully employed to derive information on the quaternary structure of a large number of soluble protein molecules, ranging in molecular mass from somewhat less than 100 kDa to several million daltons (MDa). Some examples from our own published data (Harris *et al.*, 1993a) of the high molecular mass (approx. 8 MDa) KLH1 didecamer negatively stained with 5% ammonium molybdate–1% trehalose are shown in *Figure 5.31*. For this figure, the specimen material was studied at ambient temperature and also with the specimen cooled to –175 °C. It is indicated directly that there is considerable structural protection provided by the low temperature. The quantitative splitting of the KLH1 didecamers shown here into two asymmetrical decamers can be achieved (*Figure 5.32*); this reveals the pentameric nature (D5 symmetry) of the molecule more clearly than the didecamer does. Another giant protein molecule of interest to biochemists is the approximately 4 MDa extracellular annelid haemoglobin/erythrocruorin/chlorocruorin, which was one of the first macromolecules to be studied by negative staining in the 1960s, by the eminent French haematologist and electron microscopist, Marcel Bessis. Studies on this molecule continue to the present day, by negative staining, cryonegative staining and cryoelectron microscopy of unstained frozen-hydrated specimens. The haemoglobin from the marine worm *Nereis virens* is shown in *Figure 5.33*, negatively stained with 5% ammonium molybdate–1% trehalose. This example is of particular interest because it shows both intact bilayer hexagonal molecules and molecules that are partly dissociated, due to slow instability during sample storage rather than negative staining. The loss of one or more triangular groups of subunits can be defined, together with the appearance of these smaller particles alongside the larger molecules. In this way, negative staining might be considered to provide dynamic information at the level of *macromolecular microdissection*. It is appropriate to perform a direct comparison of *Figure 5.33* with the recently published cryoelectron microscopic data on the erythrocruorin from the polychaete annelid *Eudistylia vancouverii* (de Haas *et al.*, 1996) and earthworm (*Lumbricus terrestris*) haemoglobin (Schatz *et al.*, 1995), and when the latter is imaged at low temperature in ammonium molybdate–glucose, as shown in the 15 Å resolution 3-D reconstruction in *Figure 9.1*.

The *E. coli* chaperonin GroEL (also termed cpn60) is a cylindrical molecule of molecular mass 840 kDa, constructed from two rings of seven subunits. GroEL and the smaller 70 kDa co-chaperonin GroES (cpn10) which also has a seven subunit ring-like structure, have received considerable attention from electron microscopists using negative staining (Harris *et al.*, 1994, 1995c; Langer *et al.*, 1992; Zahn *et al.*, 1993) and cryoelectron microscopy (Chen *et al.*, 1994). Although the X-ray crystal structure of these two proteins is now available, it has yet to be determined for the

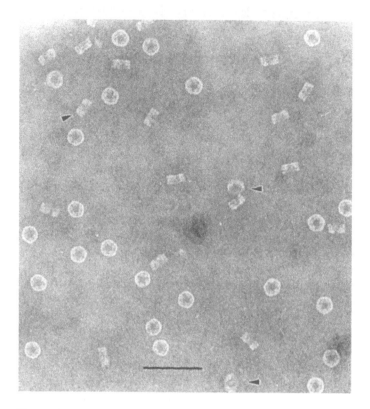

Figure 5.32:   Negatively stained KLH1 decamers.

KLH1 decamers, produced by increasing the pH of the stabilizing buffer of a KLH1 didecamer suspension from pH 7.4 to 8.5. Negatively stained with 5% ammonium molybdate containing 1% trehalose (pH 7.0). Note the pentagonal symmetry, the clarity of the five central 'collar' elements in the end-on molecules and the three 'tiers' of the side-on molecules. Some molecules are orientated at angles in between the two predominant orientations (arrows); such tilted image profiles are of importance for the recovery of 3-D image information by single particle averaging. (Cf. *Figures 5.31, 8.8, 8.9* and *10.1.*) The scale bar indicates 100 nm.

asymmetrical 'bullet-shaped' GroEL–GroES complex or the symmetrical 'American football-shaped' GroES–GroEL–GroES complex (see *Figure 5.34*). Indeed, it would appear that electron microscopy may continue to have a contribution to make, in view of the binding of a large number of different denatured/unfolded 'substrate' proteins to GroEL (Braig *et al.*, 1993; Marco *et al.*, 1994; Tsuprun *et al.*, 1995) and its ability to form an elongated filamentous form and a paracrystalline array (Harris *et al.*, 1994, 1995c). It is pertinent to note that in shallow negative stain, the symmetrical complex will almost always be orientated on its side, producing the banded ellipsoidal projection image, whereas in deeper stain some embedded molecules are able to position themselves on the more pointed end of the ellipsoid (*Figure 5.34a* and *b*).

The iron storage protein ferritin and its empty shell, apoferritin (approx.

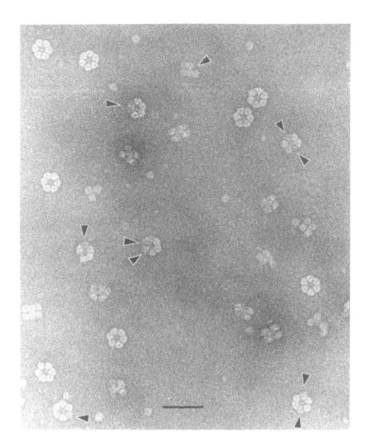

Figure 5.33: Storage instability of negatively stained haemoglobin from *Nereis virens*.
Haemoglobin from the marine annelid *N. virens*, following storage at 4°C for several weeks, negatively stained with 5% ammonium molybdate containing 1% trehalose (pH 7.0). The hexagonal molecules exhibit considerable evidence of instability, due to storage, with the loss of one or more triangular subunit groups (arrows). (Cf. *Figure 9.1*.) The scale bar indicates 50 nm.

480 kDa), have received the attention of electron microscopists for many years (see Harris, 1988; Massover, 1993). This approximately spherical protein is composed of 24 subunits and has a hollow centre occupied by an iron hydroxide crystallite core of varying size (see also Chapter 8). The variation in ferritin core size and density can be revealed by negative staining with a low density (i.e. light-atom) negative stain such as sodium tetraborate, as shown in *Figure 5.35*. Apoferritin will always generate a ring-like projection image, because it is orientation-independent to a first approximation (Harris, 1982). This point is emphasized by the ring-like projection images of the protein torin (Harris, 1969) and the *Limulus polyphemus* lectin (limulin/C-reactive protein) which also generate characteristic doublet 'side-view' projection images (see *Figure 5.36a,b*; Fernández-Morán *et al.*, 1968).

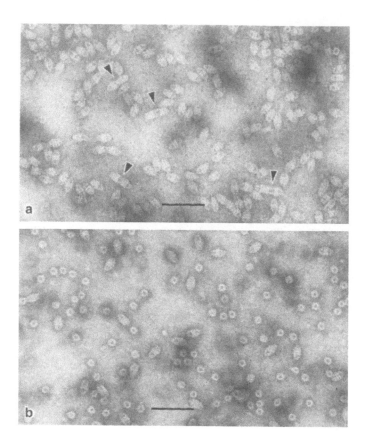

Figure 5.34: Negatively stained complexes of the *E. coli* cpn60 chaperonin GroEL with the co-chaperonin GroES.

Symmetrical (American football-shaped) complexes of the *E. coli* cpn60 chaperonin GroEL with the co-chaperonin GroES, negatively stained with 2% uranyl acetate. In (a) most of the elipsoidal complexes are orientated 'on-side'; some doublets are present (arrows). In (b) there is deeper negative stain, and the GroES–GroEL–GroES complexes are orientated 'on-side' and 'on-end'. The scale bars indicate 100 nm. Modified from Harris *et al.* (1994).

The availability of monoclonal antibodies to discrete epitopes on macromolecules has provided a highly specific mechanism for the analysis of protein structure, to which negative staining has already made a significant contribution (Boisset *et al.*, 1988; Delain *et al.*, 1988; de Haas *et al.*, 1994; Harris, 1996; Lamy *et al.*, 1993). Because of the undoubted increasing importance of this immunological approach several related examples of immunonegative staining will now be given. A monoclonal IgG antibody specific to the domain pair *fg* of the elongated multi-domain subunit of keyhole limpet haemocyanin type 2 (KLH2) has been found to generate a predominantly linear molecular arrangement within the immune complexes (see *Figure 5.37*) whereas another monoclonal IgG antibody directed

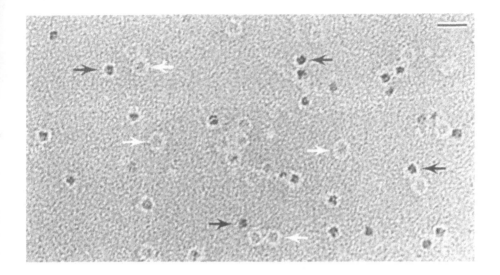

Figure 5.35: Low density negative staining of ferritin.
Horse spleen ferritin negatively stained with 2% sodium tetraborate. With this *low density* negative stain the iron hydroxide cores of the spherical molecules are revealed more clearly (black arrows) than with higher density negative stains. Also, the identification of the population of naturally occurring apoferritin (empty) molecules (white arrows) is convincingly demonstrated. The scale bar indicates 200 nm. Micrograph courtesy of Bill Massover. Reprinted from Massover (1993) Ultrastructure of ferritin and apoferritin: a review. *Micron* 24, 389–437, with permission from Elsevier Science Ltd.

to an epitope on domain a generated a totally different linkage pattern (Harris *et al.*, 1993a). From this, and other evidence of a biochemical and immunological nature, it was proposed that the 10 subunits within the intact KLH2 decamer are organized in a parallel manner. A somewhat similar linear molecular arrangement within the immune complexes of the *Bordetella pertussis* cpn60 chaperonin was found by Cejka *et al.* (1993). Image processing of contour averaged pairs supported the well defined molecular spacing and localization of the IgG molecules (*Figure 5.38*). These workers also showed the binding of Fab' fragments, which do not create immune complexes, again with the use of single particle image averaging to usefully extend the direct visual interpretation. Martin *et al.* (1994) also used this approach in their study using two different monoclonal antibodies against the *E. coli* chaperonin GroEL within which they selected groups of two or three GroEL molecules linked by IgG for image processing. In this instance the negatively stained specimens were studied by dark-field STEM imaging, yielding the data shown in *Figure 5.39*.

Very often the accurate study of IgG, Fab or Fab' attachment to a single molecule can be fraught with difficulties, but when small groups of molecules are linked with an IgG or Fab of known specificity, useful interpretations can be made regarding epitope location, even without image

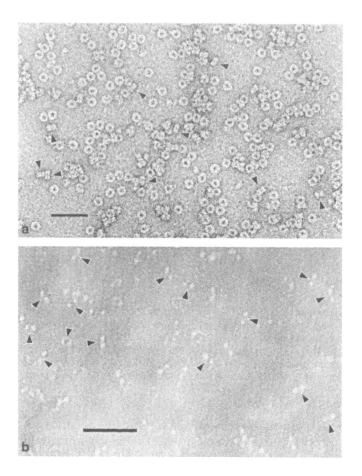

Figure 5.36: Negatively stained 'ring-like' proteins.

Examples of the ring-like proteins (a) limulin (the *Limulus polyphemus* lectin/C-reactive protein) and (b) human erythrocyte torin (Harris, 1969), negatively stained with 2% uranyl acetate and 2% sodium phosphotungstate (pH 7.0), respectively. Arrows indicate the doublet 'side-on' images, which predominate in (b) because of increased molecular freedom due to reduced adsorption by the carbon support film in the neutral pH anionic negative stain. The scale bars indicate 50 nm.

processing. Indeed, the image variability may preclude such an approach whereas the visual interpretation can more readily encompass a number of variations, all of which may be compatible with the final structural interpretation. Because of the presence of multiple epitope sites within oligomeric complexes, the quantity of bound IgG and the number of possible linkage patterns can be fairly extensive. Nevertheless, some linkage patterns will probably predominate and firm conclusions can then be drawn. This was indeed the approach of Kopp *et al.* (1993, 1995) who studied the location of several individual subunits within the human 20S

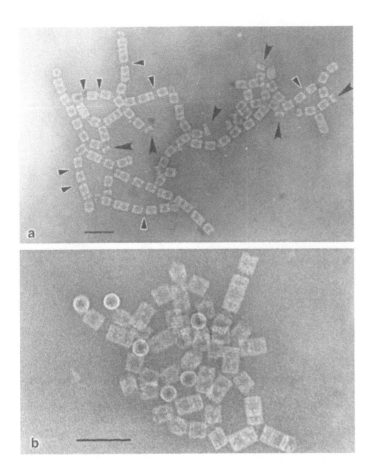

Figure 5.37: Negatively stained immune complexes of keyhole limpet haemocyanin type 2 (KLH2).

A didecamer-enriched fraction of KLH2 following interaction with a monoclonal antibody directed against the subunit domain pair *fg* and separation of immune complexes by gel filtration chromatography (see Harris *et al.*, 1993a). Note in (a) the predominant linear end-to-end linkage of the KLH2 by multiple IgG molecules (small arrows), with limited branching points and no side-to-side molecular linkages. Also note, the termination of several of the linear chains by KLH2 decamers (large arrows). In (b) a more compact 3-D immune complex indicates the loss of detail when molecular superimposition occurs. The antibody linkage pattern in (b) does, nevertheless, agree with that shown in (a). Negatively stained with 5% ammonium molybdate containing 1% trehalose (pH 7.0). The scale bars indicate 100 nm.

proteasome complex (*Figures 5.40, 5.41* and *5.42*). Somewhat similar studies were performed by Grziwa *et al.* (1991, 1992) using the *Thermoplasma acidophilum* 20S proteasome, which contains only two types of subunit.

It is of importance for the overall discussion to note that the immune complexes shown in *Figures 5.37–5.42* have been negatively stained with ammonium molybdate, uranyl acetate, uranyl formate and sodium phos-

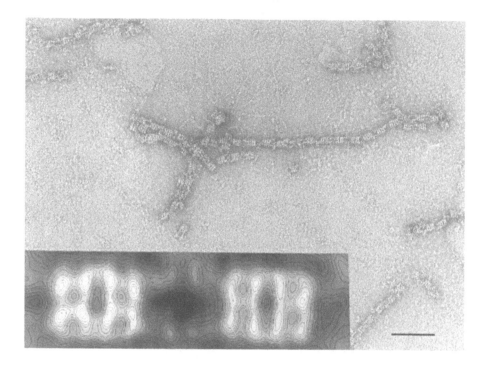

Figure 5.38: Negatively stained images of cpn60 from *Bordetella pertussis* after interaction with monoclonal antibody 54G8.

The molecules, mostly in side view, are connected by the Fab fragments of the antibody molecule and form characteristically long chains. Negatively stained with 1% uranyl acetate. The inset shows a contoured average of pairs of molecules cut out from a chain. Antibody molecules extend from the four corners of the molecule. The two antibodies linking the molecules are clearly seen with two strong centres of mass midway between the two cpn60 molecules, interpreted to be the Fab fragments. The scale bar indicates 100 nm. Micrograph courtesy of Zdenka Cejka and Martin Kessel. Reproduced from Cejka *et al.* (1993) Localization of the binding site of an antibody affecting ATPase activity of chaperonin cpn60 from *Bordetella pertussis*. *J. Struct. Biol.* 111, 34–38, with permission from Academic Press.

photungstate. All four negative stains proved successful for the localization of the bound antibody molecules and definition of the linkage patterns. Recent experience of mine using very high molecular mass molluscan haemocyanin (KLH1 and KLH2) favours the use of ammonium molybdate–trehalose (Harris *et al.*, 1995a), but there is no question that valid meaningful data can be obtained from immune complexes revealed by almost any of the available negative stains (stain-induced protein aggregation or dissociation must always be avoided).

The production of negatively stained 2-D crystals of membrane proteins, and their importance for the development of electron crystallographic studies has been mentioned above (see *Figures 5.11, 5.17* and *5.18*). A number of examples of 2-D crystals produced on mica by the NS–CF technique will now be given, leading then to the study of small 3-D crystals by negative staining.

One of the most significant benefits to be derived from the NS–CF technique is the facility to detect early stages of the 2-D crystallization process (nucleation) and the variety of 2-D crystals and pre-crystalline partly ordered 2-D forms that can be produced. This enables the necessary crystallization conditions to be narrowed down, during trial experiments with any previously unstudied protein (see Chapter 3.5.2). *Figure 5.43* shows the soluble 16S Mg–ATPase complex from *Xenopus* oocytes (Peters *et al.*, 1992) where small hexagonal 2-D crystals have formed. In this instance an insufficiency of protein molecules has prevented the creation of larger,

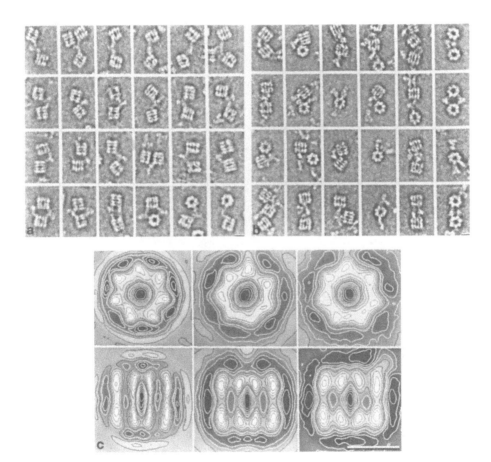

Figure 5.39: STEM dark-field images of GroEL–antibody complexes, negatively stained with uranyl formate.
(a) Complexes of GroEL–anti C40 antibodies and (b) complexes of GroEL–anti N15 antibodies. (c) Projection averages of GroEL and GroEL–antibody complexes. Left column: GroEL tetradecamers without bound antibodies; middle column: GroEL–anti N15 antibody complexes; right column: GroEL–anti C40 antibody complexes. The scale bars indicate 10 nm. Micrographs courtesy of Andreas Engel. Reproduced from Martin *et al.* (1994) Topology of the morphological domains of the chaperonin GroEL visualized by immuno-electron microscopy. *Biol. Chem. Hoppe Seyler* 375, 635–639, with permission from Walter de Gruyter & Co.

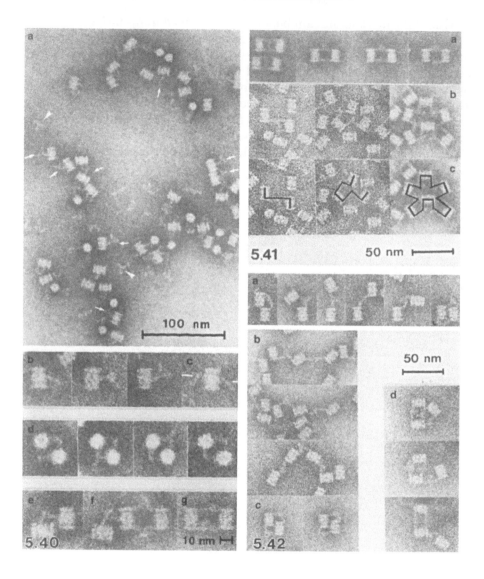

Figures 5.40–5.42: Immune complexes of human 20S proteasome.

Examples of immune complexes produced by a range of different monoclonal antibodies against human placental 20S proteasome subunits (from Kopp *et al.*, 1993 , 1995; refer to these publications and to Kristensen *et al.*, 1994, for full information), negatively stained with sodium phosphotungstate (pH 7.2). *Figure 5.40*: immune complexes produced with monoclonal antibody MCP444, directed against a central α-type subunit HsN3. *Figure 5.41*: immune complexes produced with monoclonal antibody MCP34 (XAPC-7) specific for α-type subunit 12. *Figure 5.42*: immune complexes produced using monoclonal antibody MCP20 (HC2) specific for α-type subunit 7. Micrographs courtesy of Friedrich Kopp. *Figure 5.40* reprinted from Kopp *et al.* (1995) The human proteasome subunit HsN3 is located in the inner rings of the complex dimer. *J. Mol. Biol.* 248, 264–272, with permission from Academic Press Ltd. *Figure 5.41* and *5.42* reprinted from Kopp *et al.* (1993) Evidence indicating that the human proteasome is a complex dimer *J. Mol. Biol.* 229, 14–19, with permission from Academic Press Ltd.

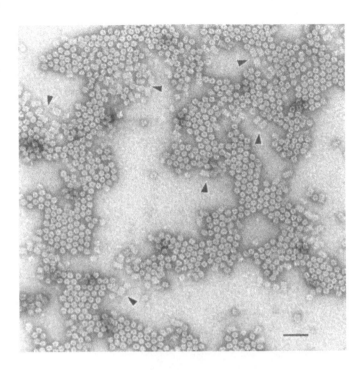

Figure 5.43: 2-D crystals of the 16S Mg–ATPase complex from *Xenopus* oocytes.
The creation of small hexagonal 2-D crystals from the 16S soluble Mg–ATPase complex from *Xenopus* oocytes (Peters *et al.*, 1992), produced on mica by the NS–CF technique at pH 8.0, with uranyl acetate as the negative stain. In this instance, 2-D crystal formation, with molecules orientated 'end-on', is interfered with by randomly positioned molecules adopting a 'side-on' orientation (arrows). The scale bar indicates 50 nm.

more continuous, 2-D crystals. But the tendency of the molecule to alternate between the two favoured orientations (end-on/side-on) may also interfere with crystal formation. Much larger 2-D crystals can be formed readily from human erythrocyte catalase (*Figure 5.44*) and meaningful 2-D projection averages have been obtained from them (see also Chapter 10). Human erythrocyte catalase is particularly interesting because of its ability to generate several different 2-D crystal forms and ordered 2-D arrays (Harris and Holzenburgh, 1989, 1995; Harris *et al.*, 1993b); catalase has been a molecule of long-standing interest to electron microscopists, as a well characterized enzyme, upon which a range of different technical procedures can be performed (Erickson and Klug, 1971; Schröder, 1992) and in the form of thin 3-D crystals as a calibration standard for instrumental magnification (Wrigley, 1968).

The didecamer of keyhole limpet haemocyanin type 1 (KLH1) readily generates 2-D crystals (*Figure 5.45*), from which a low resolution (approx. 27 Å) 2-D projection image reconstruction has been obtained (Harris *et al.*, 1992). The generation of larger, well ordered 2-D crystals from this

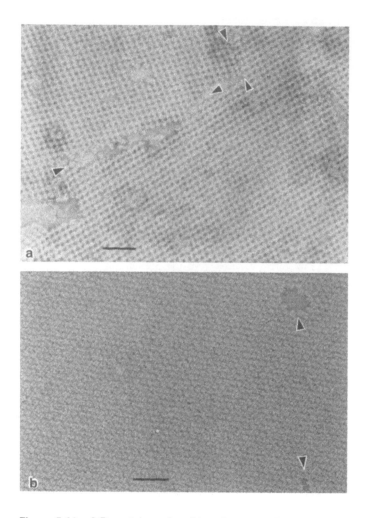

Figure 5.44: 2-D crystals produced from human erythrocyte catalase.

Examples of two different 2-D crystals produced from human erythrocyte catalase by the NS–CF technique (Harris and Holzenburg, 1989; Harris *et al.*, 1993b), negatively stained with 2% uranyl acetate. Note the crystal discontinuity between arrows (a) and the absence of molecules within the 2-D crystal lattice (arrows, b). The scale bars indicate 50 nm.

protein, has proved to be difficult. Thus, no attempts have been made to-date to produce tilt series images, for merging and production of a crystallographic 3-D reconstruction, apart from the early and rather significant negative staining study of Mellema and Klug (1972) with a different molluscan haemocyanin. Instead, 3-D information has been recovered through single particle image processing (Dube *et al.*, 1995; *Figures 5.31, 8.8, 8.9* and *10.1*).

2-D crystals of the archeobacterium *Thermoplasma acidophilum* 20S proteasome have also proved to be difficult to produce by the NS–CF technique, in this instance because of the marked rotational variation of molecules within the 2-D array (Kleinz *et al.*, 1992). Nevertheless, some success has been achieved (*Figure 5.46*). From this 2-D crystal, crystallographic image processing was able to generate a meaningful projection average, apparently with six molecules in the unit cell. The 20S proteasome also has the ability to generate small 3-D crystals on mica, when the crystallization procedure is performed in the presence of PEG alone (*Figure 5.47*), as demonstrated by Harris *et al.* (1992).

The use of negative staining to study small 3-D crystals produced by other crystallization procedures has not been widely applied. However, the study of 3-D crystals of the enzyme alcohol oxidase from *Hansenula polymorpha* by Vonck and van Bruggen (1992) is a good example (*Figure 5.48*). Unfortunately, the recovery of 2-D and 3-D information from such

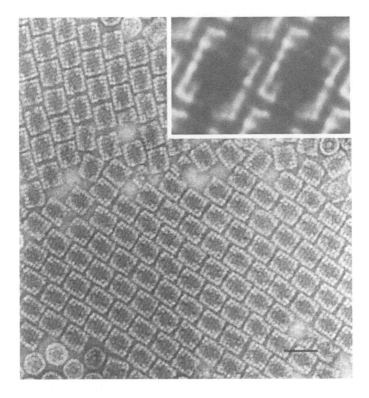

Figure 5.45: 2-D crystals of keyhole limpet haemocyanin didecamers.
    Examples of 2-D crystals of keyhole limpet haemocyanin didecamers produced by the NS–CF techniqe and negatively stained with 2% uranyl acetate (modified from Harris *et al.*, 1992). The production of large 2-D crystals, even from highly purified KLH1 didecamers, has proved to be technically difficult. The inset shows the 2-D projection average, which retains the symmetry features of the individual decamers, created by the precise angular orientation of each molecule within the 2-D crystal. (Cf *Figures 5.31, 8.8, 8.9* and *10.1*.) The scale bar indicates 50 nm.

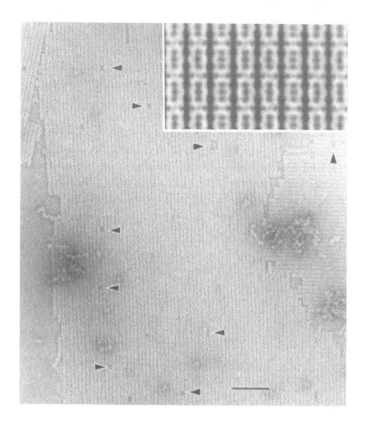

Figure 5.46: A 2-D crystal of the 20S proteasome from *Thermoplasma acidophilum*.
A 2-D crystal of the 20S proteasome from *T. acidophilum*, produced by the NS–CF procedure. Because of considerable rotational variation of individual molecules within the crystal (Kleinz *et al.*, 1992) crystallographic image processing (inset) suggests the presence of six molecules within the unit cell, rather than the more usual two or four. Note the absence of individual and small numbers of molecules at places within the 2-D crystal lattice (arrows). (Data produced by JRH, when working within the laboratory of Wolfgang Baumeister.) The scale bar indicates 100 nm.

specimens by crystallographic image processing cannot easily be pursued, usually because of the difficulties introduced due to overlapping molecules in consecutive molecular lattices and the inherent complexity of such analysis. Most molecules that form small 3-D crystals will also produce larger crystals, for example the *T. acidophilum* 20S proteasome (Löwe *et al.*, 1995) which are suitable for higher resolution X-ray diffraction studies. As 3-D crystals have not yet been produced from the 26S proteasome complex (Peters *et al.*, 1993), this macromolecule remains a suitable target for further structural studies by TEM, directed towards the production of a 3-D reconstruction.

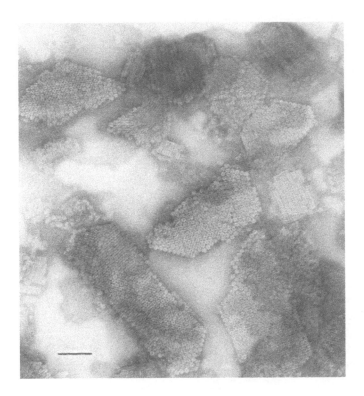

Figure 5.47: 3-D crystals of 20S proteasome produced by the NS–CF procedure.
Small, partly formed 3-D crystals produced from the 20S proteasome of *T. acidophilum* produced by the NS–CF procedure using PEG alone (Harris *et al.*, 1992). Negatively stained with 2% uranyl acetate. (Cf. *Figure 5.30.*) The scale bar indicates 100 nm.

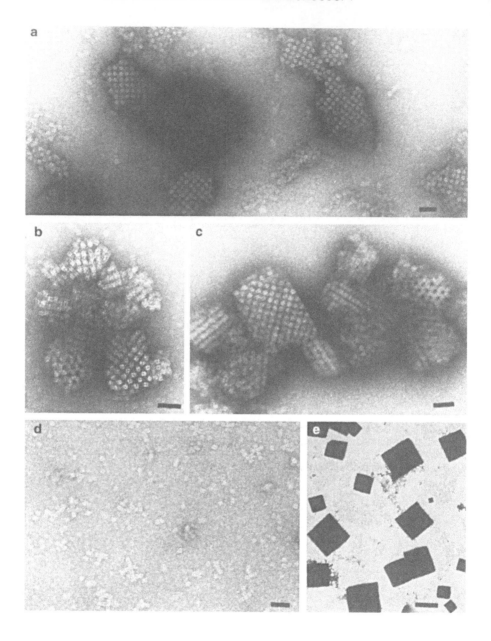

Figure 5.48: 3-D crystals of alcohol oxidase from *Hansenula polymorpha*.

Small 3-D crystals of alcohol oxidase from *H. polymorpha*, grown in solution. Negatively stained with 1% uranyl acetate. The scale bars indicate 50 nm (a–d) and 1 μm (e). Micrographs courtesy of Janet Vonck. Reproduced from Vonck and van Bruggen (1992) Architecture of peroxisomal alcohol oxidase crystals from the methylotropic yeast *Hansenula polymorpha* as deduced by electron microscopy. *J. Bacteriol.* 174, 5391–5399, with permission from the American Society for Microbiology.

References

Abrahams JP, Leslie AGW, Lutter R, Walker JE. (1994) Structure at 2.8 Å of F_1–ATPase from bovine heart mitochondria. *Nature* **370**, 621–628.

Albiges-Rizo C, Barge A, Ruigrok RWH, Timmins PA, Chroboczek J. (1991) Human adenovirus serotype 3 fiber protein. *J. Biol. Chem.* **266**, 3961–3967.

Amos LA. (1982) Tubulin and associated proteins. In *Electron Microscopy of Proteins* (ed. JR Harris) Vol. 3. Academic Press, London, pp. 207–250.

Baumeister W, Engelhardt H. (1987) Three-dimensional structure of bacterial surface layers. In *Electron Microscopy of Proteins* (eds JR Harris, RW Horne) Vol. 6. Academic Press, London, pp. 109–154.

Benedetti EL, Emmelot P. (1968) Hexagonal array of subunits in tight junctions separated from rat liver plasma membranes. *J. Cell Biol.* **38**, 15–24.

Beveridge TJ, Graham LL. (1991) Surface layers of bacteria. *Microbiol. Rev.* **55**, 684–705.

Bhakdi S, Tranum-Jensen J, Sziegloiet A. (1985) Mechanism of damage by streptolysin-O. *Infect. Immun.* **47**, 52–60.

Bhakdi S, Weller U, Walev I, Martin E, Jonas D, Palmer M. (1993) A guide to the use of pore-forming toxins for controlled permeabilization of cell membranes. *Med. Microbiol. Immunol.* **182**, 167–175.

Boekema EJ, Berden JA, van Heel MG. (1986) Structure of the mitochondrial F_1–ATPase studied by electron microscopy and image processing. *Biochim. Biophys. Acta* **851**, 353–360.

Boekema EJ, van Heel M, Gräber P. (1988) Structure of the ATP synthase from chloroplasts studied by electron microscopy and image processing. *Biochim. Biophys. Acta* **933**, 365–371.

Boekema EJ, Dekker JP, Rögner M, Witt HT, van Heel M. (1989) Refined analysis of the trimeric structure of the isolated photosystem I complex from the thermophilic cyanobacterium *Synechococcus sp. Biochim. Biophys. Acta* **974**, 81–87.

Boekema EJ, Boonstra AF, Dekker JP, Rögner M. (1994) Electron microscopic structural analysis of photosystem I, photosystem II, and the cytochrome b6/f complex from green plants and cyanobacteria. *J. Bioenerg. Biomemb.* **26**, 17–29.

Boissset N, Taveau J-C, van Leuven F, Delain E, Lamy JN. (1988) Image analysis and three-dimensional model of chymotrypsin-transformed alpha 2-macroglobulin complexed with a monoclonal antibody specific for this conformation. *Biol. Cell* **64**, 45–55.

Boudin M-L, Moncany M, d'Halluin J-C, Boulanger PA. (1979) Isolation and characterization of adenovirus type 2 vertex capsomer (penton base). *Virology* **92**, 125–138.

Braig K, Simon M, Furuya F, Hainfeld JF, Horwich AL. (1993) A polypeptide bound by the chaperonin GroEL is localized within a central cavity. *Biochemistry* **90**, 3978–3982.

Bremner A, Henn C, Goldie KN, Engel A, Smith PR, Aebi U. (1994) Towards atomic interpretation of F-actin filament three-dimensional reconstructions. *J. Mol. Biol.* **242**, 683–700.

Bullit E, Makowski L. (1995) Structural polymorphism of baterial adhesion pili. *Nature* **373**, 164–167.

Castón JR, Berenguer J, de Pedro MA, Carrascosa JL. (1993) S-layer protein from *Thermus thermophilus* HB8 assembles into porin-like structures. *Molec. Microbiol.* **9**, 65–75.

Cejka Z, Gould-Kostka J, Burns D, Kessel M. (1993) Localization of the binding site of an antibody affecting ATPase activity of chaperonin cpn60 from *Bordetella pertussis*. *J. Struct. Biol.* **111**, 34–38.

Chapman JA, Tzaphlidou M, Meek KM, Kadler KE. (1990) The collagen fibril – a model system for studying staining and fixation of a protein. *Electr. Microsc. Rev.* **3**, 143–182.

Chen S, Roseman AM, Hunter AS, Wood SP, Burson SG, Ranson NA, Clarke AR, Saibil HR. (1994) Location of a folding protein and shape changes in GroEL–GroES complexes imaged by cryo-electron microscopy. *Nature* **371**, 261–264.

Delain E, Barray M, Tapon-Bretaudiere J, Pochon F, Marynen P, Cassiman J-J, van den Berghe H, van Leuven F. (1988) The molecular organization of human α_2-macroglobulin. *J. Biol. Chem.* **263**, 2981–2989.

Dube P, Orlova EV, Zemlin F, van Heel M, Harris JR, Markl J. (1995) Three-dimensional structure of keyhole limpet hemocyanin (KLKH) by cryoelectron microscopy and angular reconstitution. *J. Struct. Biol.* **115**, 226–232.

Dubochet J, Adrian M, Richter K, Garces J, Wittek R. (1994) Structure of intracellular mature vaccinia virus observed by cryoelectron microscopy. *J. Virol.* **68**, 1935–1941.

Duncan JL, Schlegel R. (1975) Effects of streptolysin O on erythrocyte membranes, liposomes and lipid dispersions. *J. Cell Biol.* **67**, 160–173.

Dürr R, Hegerl R, Volker S, Santarius U, Baumeister W. (1991) Three-dimensional reconstruction of the surface protein of *Pyrodictium brockii*: comparing two image processing strategies. *J. Struct. Biol.* **106**, 181–190.

Egelman EH, Orlova A. (1995) Allostery, cooperativity, and different structural states in F-actin. *J. Struct. Biol.* **115**, 159–162.

Erickson H, Klug A. (1971) Measurement and compensation of defocusing and aberrations by Fourier processing of electron micrographs. *Phil. Trans. R. Soc. Lond.* **B 261**, 105–118.

Fernández-Morán H, Marchalonis JJ, Edelman GM. (1968) Electron microscopy of a hemagglutinin from *Limulus polyphemus*. *J. Mol. Biol.* **32**, 467–469.

Fok AK, Wang H, Katayama A, Aihara MS, Allen RD. (1994) 22S axonemal dynein is preassembled and functional prior to being transported to and attached on the axonemes. *Cell Motil. Cytoskel.* **29**, 215–224.

Ford RC, Rosenberg MF, Shepherd FH, McPhie PL, Holzenburg A. (1995) Photosystem II 3-D structure and the role of the extrinsic subunits in photosynthetic oxygen evolution. *Micron* **26**, 133–140.

Fujiyama Y, Yokoyama K, Yoshida M, Wakabayashi T. (1990) Electron microscopy of the reconstituted complexes of the F_1–ATPase with various subunit constitution revealed the location of the gamma subunit in the central cavity of the molecule. *FEBS Lett.* **271**, 111–114.

Ghoshroy S, Goodenough DA, Sosinsky GE. (1995) Preparation, characterization, and structure of half gap junctional layers split with urea and EGTA. *J. Memb. Biol.* **146**, 15–28.

Gouaux JE, Braha O, Hobaugh MR, Song L, Cheley S, Shustak C, Baley H. (1994) Subunit stoichiometry of staphyolococcal α-hemolysin in crystals and on membranes: a heptameric transmembrane pore. *Proc. Natl Acad. Sci. USA* **91**, 12828–12831.

Grziwa A, Baumeister W, Dahlmann B, Kopp F. (1991) Localization of subunits in proteasomes from *Thermoplasma acidophilum* by immunoelectron microscopy. *FEBS Lett.* **290**, 186–190.

Grziwa A, Dahlamnn B, Cejka Z, Santarius U, Baumeister W. (1992) Localization of a sequence motif complementary to the nuclear localization signal in proteasomes from *Thermoplasma acidophilum*. *J. Struct. Biol.* **109**, 168–175.

Guan T, Müller S, Klier G, Paanté N, Blevbitt JM, Haner M, Paschal B, Aebi U, Gerace L. (1995) Structural analysis of the p62 complex, an assembly of *O*-linked glycoproteins that localizes near the central gated channel of the nuclear pore complex. *Mol. Biol. Cell* **6**, 1591–1603.

Guo XW, Mannella CA. (1992) Classification of projection images of crystalline arrays of the mitochondrial voltage-dependent anion-selective channel embedded in aurothioglucose. *Biophys. J.* **63**, 418–427.

Guo XW, Smith PR, Cognon B, d'Arcangelis D, Dolginova E, Mannella CA. (1995) Molecular design of the voltage-dependent, anion-selective channel in the mitochondrial outer membrane. *J. Struct. Biol.* **114**, 41–59.

de Haas F, Perton FG, van Breemen JFL, Dijkema JH, Beintema JJ, van Bruggen EFJ. (1994) Identification of two antibody-interactive sites on the surface of *Panulirus interruptus* hemocyanin. *Eur. J. Biochem.* **222**, 155–166.

de Haas F, Taveau J-C, Boisset N, Lambert O, Vinogradov SN, Lamy JN. (1996) Three-dimensional reconstruction of the chlorocruorin of the polychaete annelid *Eudistylia vancouverii. J. Mol. Biol.* **255**, 140–153.

Harris JR. (1969) Some negative contrast staining features of a protein from erythrocyte ghosts. *J. Mol. Biol.* **46**, 329–335.

Harris JR. (1978) Biochemistry and ultrastructure of the nuclear envelope. *Biochim. Biophys. Acta* **515**, 55–104.

Harris JR. (1982) Nonenzymic proteins. In *Electron Microscopy of Proteins* (ed. JR Harris) Vol. 2. Academic Press, London, pp. 49–103.

Harris JR. (1988) Electron microscopy of cholesterol. *Micron Microsc. Acta* **19**, 19–32.

Harris JR. (1991) The negative staining–carbon film procedure: technical considerations and a survey of macromolecular applications. *Micron Microsc. Acta* **22**, 341–359.

Harris JR. (1996) Immunonegative staining: epitope localization on macromolecules. *Methods,* in press.

Harris JR, Agutter PS. (1970) A negative staining study of human erythrocyte ghosts and rat liver nuclear membranes. *J. Ultrastruct. Res.* **33**, 219–232.

Harris JR, Holzenburg A. (1989) Transmission electron microscopy of negatively stained human erythrocyte catalase. *Micron Microsc. Acta* **20**, 223–238.

Harris JR, Holzenburg A. (1995) Human erythrocyte catalase: 2-D crystal nucleation and multiple 2-D crystal forms. *J. Struct. Biol.* **115**, 102–112.

Harris JR, Horne RW. (1994) Negative staining: a brief assessment of current technical benefits, limitations and future possibilites. *Micron* **25**, 5–13.

Harris JR, Markl J. (1992) Electron microscopy of a double helical tubular filament in keyhole limpet (*Megathura crenulata*) haemolymph. *Cell. Tiss. Res.* **269**, 411–420.

Harris JR, Marshall P. (1981) The nuclear envelope and the nuclear pore complex: some electron microscopic and biochemical considerations. In *Electron Microscopy of Proteins* (ed. JR Harris) Vol. 1. Academic Press, London, pp. 39–124.

Harris JR, Pfeifer G, Pühler G, Baumeister W. (1992) Production of 3-D microcrystals from *Thermoplasma acidophilum* multicatalytic proteinase/proteasome by the negative staining–carbon film technique. In *Electron Microscopy*, Vol. 1. EUREM 92, Granada, Spain, pp. 383–384.

Harris JR, Gebauer W, Markl, J. (1993a) Immunoelectron microscopy of keyhole limpet hemocyanin: a parallel subunit model. *J. Struct. Biol.* **111**, 96–113.

Harris JR, Volker S, Engelhardt H, Holzenburg H. (1993b) Human erythrocyte catalase: new 2-D crystal forms and image processing. *J. Struct. Biol.* **111**, 22–33.

Harris JR, Plückthun A, Zahn R. (1994) Transmission electron microscopy of GroEL, GroES and the symmetrical GroEL/ES complex. *J. Struct. Biol.* **112**, 216–230.

Harris JR, Gebauer W, Markl J. (1995a) Keyhole limpet hemocyanin (KLH): negative staining in the presence of trehalose. *Micron* **26**, 25–33.

Harris JR, Gebauer W, Söhngen SM, Markl J. (1995b) Keyhole limpet hemocyanin (KLH): purification of intact KLH1 through selective dissociation of KLH2. *Micron* **26**, 201–212.

Harris JR, Zahn R, Plückthun A. (1995c) Electron microscopy of the GroEL-GroES filament. *J. Struct. Biol.* **115**, 68–77.

Harris JR, Gerber M, Gebauer W, Wernicke W, Markl J. (1996) Negative stains containing trehalose: application to tubular and filamentous structures. *J. Microsc. Soc. Am.* **2**, 43–52.

Hebert H, Olofsson A, Thelestam M, Skriver E. (1992) Oligomer formation of staphylococcal α-toxin analyzed by electron microscopy and image processing. *FEMS Microbiol. Immunol.* **105**, 5–12.

Heitlinger E, Peter M, Häner M, Lustig A, Aebi U, Nigg EA. (1991) Expression of chicken lamin b2 in *Escherichia coli*: characterization of its structure, assembly, and molecular interactions. *J. Cell Biol.* **113**, 485–495.

Henderson R, Baldwin JM, Ceska TA, Beekmann E, Zemlin F, Downing K. (1990) A model for the structure of bacteriorhodopsin based on high resolution electron microscopy. *J. Mol. Biol.* **213**, 899–929.

Hess M, Cuzanage A, Ruigrok RWM, Chroboczek J, Jacrot B. (1995) The avian adenovirus penton: two fibres and one base. *J. Mol. Biol.* **376**, 379–385.

Hirose K, Lockhart A, Cross RA, Amos LA. (1995) Nucleotide-dependent angular change in kinesin motor domain bound to tubulin. *Nature* **376**, 277–279.

Hoenger A, Sabalin EP, Vale RD, Fletterick RJ, Milligan RA. (1995) Three-dimensional structure of a tubulin-motor-protein-complex. *Nature* **376**, 271–274.

Hofhaus G, Weiss H, Leonard K. (1991) Electron microscopic analysis of the peripheral and membrane parts of mitochondrial NADH dehydrogenase (complex I). *J. Mol. Biol.* **221**, 1027–1043.

Holzenburg A, Shepherd FH, Ford RC. (1994) Localization of the oxygen-evolving complex of photosystem II by Fourier difference analysis. *Micron* **25**, 447–451.

Horne RW. (1986) Electron microscopy of crystalline arrays of adenoviruses and their components. In *Electron Microscopy of Proteins* (eds JR Harris, RW Horne) Vol. 5. Academic Press, London, pp. 71–101.

Horne RW, Wildy P. (1979) An historical account of the development and applications of the negative staining technique to the electron microscopy of viruses. *J. Microsc.* **117**, 103–122,

Horne RW, Pasquali Ronchetti I, Hobart JM. (1975) A negative staining–carbon film technique for studying viruses in the electron microscope. II. Application to adenovirus type 5. *J. Ultrastruct. Res.* **51**, 233–252.

Horne RW, Hobart JM, Markham R. (1976) Electron microscopy of tobacco mosaic virus prepared with the aid of negative staining–carbon film techniques. *J. Gen. Virol.* **31**, 265–269.

Horne RW, Harnden JM, Hull R. (1977) The *in vitro* crystalline formations of turnip rosette virus. I. Electron microscopy of 2-D and 3-D arrays. *Virology* **82**, 150–162.

Hyatt AD. (1991) Immunogold labelling techniques. In *Electron Microscopy in Biology: a Practical Approach* (ed. JR Harris). IRL Press, Oxford, pp. 59–81.

Kiselev NA, Shpitzberg CL, Vainshtein BK. (1967) Crystallization of catalase in the form of tubes with monomolecular walls. *J. Mol. Biol.* **25**, 433–441.

Kistler J, Bond J. Donaldson P. Engel A. (1993) Two distinct levels of gap junction assembly *in vitro. J. Struct. Biol.* **110**, 28–38.

Kistler J, Goldie K, Donaldson P, Engel A. (1994) Reconstitution of native-type noncrystalline lens fiber gap junctions from isolated hemichannels. *J. Cell Biol.* **126**, 1047–1058.

Kleinz J, Harris JR, Pfeifer G, Hegerl R, Puhler G, Baumeister W. (1992) Rotational orientation of proteasomes in single-particle and paracrystal preparations. In *Electron Microsocopy*, Vol. 1. EUREM 92, Granada, Spain, pp. 415–416.

Knutton S. (1982) Specialized membranes. In *Electron Microscopy of Proteins* (ed. JR Harris) Vol. 2. Academic Press, London, pp. 261–305.

Konig N, Zampighi GA. (1995) Purification of bovine lens cell-to-cell channels composed of connexin44 and connexin50. *J. Cell Sci.* **108**, 3091–3098.

Kopp F, Dahlmann B, Hendil KB. (1993) Evidence indicating that the human proteasome is a complex dimer. *J. Mol. Biol.* **229**, 14–19.

Kopp F, Kristensen P, Hendil KB, Johnsen A, Sobek A, Dahlman B. (1995) The human proteasome subunit HsN3 is located in the inner rings of the complex dimer. *J. Mol. Biol.* **248**, 264–272.

Kristensen P, Johnsen AH, Uerkvitz W, Tanaka K, Hendil KB. (1994). Human proteasome subunits from 2-D gels identified by partial sequencing. *Biochem. Biophys. Res. Commun.* **205**, 1785–1789.

Kühlbrandt W, Wang DN, Fujiyoshi X. (1994) Atomic model of plant light harvesting complex by electron crystallograpy. *Nature* **367**, 614–621.

Lamy J, Gielens C, Lambert O, Taveau JC, Motta G, Loncke P, de Geest N, Preaux G, Lamy J. (1993) Further approaches to the quaternary structure of *Octopus* hemocyanin: a model based on immunoelectron microscopy and image processing. *Arch. Biochem. Biophys.* **305**, 17–19.

Langer T, Pfeifer G, Martin J, Baumeister W, Hartl F-U. (1992) Chaperonin-mediated protein folding: GroES binds to one end of the GroEL cylinder, which accommodates the protein substrate within its central cavity. *EMBO J.* **11**, 4757–4765.

Lehman W, Craig R, Vibert P. (1994) Ca^{2+}-induced tropomyosin movement in *Limulus* thin filaments revealed by three-dimensional reconstruction. *Nature* **368**, 65–67.

Löwe J, Stock D, Jap B, Zwickl P, Baumeister W, Huber R. (1995) Crystal structure of the 20S proteasome from the archeon *T. acidophilum* at 3.4 Å resolution. *Science* **268**, 533–539.

Madeley CR, Carter MJ. (1986) The reoviruses. In *Electron Microscopy of Proteins* (eds JR Harris, RW Horne) Vol. 5. Academic Press, London, pp. 259–291.

Marco S, Carrascosa JL, Valpuesta JM. (1994) Reversible interaction of β-actin along the channel of the TCP-1 cytoplasmic chaperonin. *Biophys. J.* **67**, 364–368.

Marshall P, Harris JR. (1979) Isolation of nuclear pore complexes from Triton X-100 extracted rat liver nuclear envelope. *Biochem. Soc. Trans.* **7**, 928–929.

Martin J, Goldie KN, Engel A, Hartl F-U. (1994) Topology of the morphological domains of the chaperonin GroEL visualized by immuno-electron microscopy. *Biol. Chem. Hoppe Seyler* **375**, 635–639.

Massover WH. (1993) Ultrastructure of ferritin and apoferritin: a review. *Micron* **24**, 389–437.

McDade WA, Josephs R. (1993) On the formation and crystallization of sickle haemoglobin microfibres. *J. Struct. Biol.* **110**, 90–97.

McDermott G, Prince SM, Freer AA, Horthornthwaite-Lawless AM, Papiz MZ, Cogdell RJ, Isaacs MW. (1995) Crystal structure of an integral membrane light-harvesting complex from photosynthetic bacteria. *Nature* **374**, 517–521.

Mellema JE, Klug, A. (1972) Quaternary structure of gastropod haemocyanin. *Nature* **229**, 146–150.

Misel DL. (1978) Image enhancement and interpretation. In *Practical Methods in Electron Microscopy* (ed. AM Glauert) Vol. 7. Elsevier/North-Holland Biomedical Press, Amsterdam, pp. 188–184.

Nermut MV. (1991) Unorthodox methods of negative staining. *Micron Microsc. Acta* **22**, 327–339.

Nermut MV, Perkins WJ. (1979) Consideration of the three-dimensional structure of the adenovirus hexon from electron microscopy and computer modelling. *Micron* **10**, 247–266.

Norcum MT, Wilkinson DA, Carlson C, Hainfel JF, Carlson GM. (1994) Structure of phosphorylase kinase: three-dimensional model derived from stained and unstained electron micrographs. *J. Mol. Biol.* **241**, 94–102.

Parker MW, Buckley JT, Postma JPM, Tucker AD, Leonard K, Pattus F, Tsernoglou D. (1994) Structure of the *Aeromonas* toxin proaerolysin in its water-soluble and membrane-channel states. *Nature* **367**, 292–295.

Peters J-M, Harris JR, Lustig A, Muller S, Engel A, Volker S, Franke WW. (1992) The ubiquitous soluble Mg^{2+}–ATPase complex: a structural study. *J. Mol. Biol.* **223**, 557–571.

Peters J-M, Cejka Z, Harris JR, Kleinschmidt JA, Baumeister W. (1993) Structural features of the 26S proteasome complex. *J. Mol. Biol.* **234**, 932–937.

Petterson U, Höglund S. (1969) Structural proteins of adenoviruses III. Purification and characterization of the adenovirus type 2 penton antigen. *Virology* **39**, 90–106.

Pum W, Weinhandl M, Hödl C, Sleytr UB. (1993) Large-scale recrystallization of the S-layer of *Bacillus coagulans* E38-66 at the air/water interface and on lipid films. *J. Bacteriol.* **175**, 2762–2766.

Ruigrok RWH, Barghe A, Albiges-Rizo C, Dayan S. (1990) Structure of adenovirus fibre II. Morphology of single fibres. *J. Mol. Biol.* **215**, 589–596.

Ruiz T, Rank J-L, Diaz-Avalos R, Caspar DL, DeRosier DJ. (1994) Electron diffraction of helical particles. *Ultramicroscopy* **55**, 383–395.

Sara M, Sleytr UB. (1996) Biotechnology and biomimetic with crystalline bacterial cell surface layers (S-layers). *Micron*, in press.

Schatz M, Orlova EV, Dube P, Jäger J, van Heel M. (1995) Structure of *Lumbricus terrestris* hemoglobin at 30 Å resolution determined using angular reconstitution. *J. Struct. Biol.* **114**, 28–40.

Schröder RR. (1992) Zero-loss energy filtered imaging of frozen-hydrated proteins: model calculations and implications for future developments. *J. Microsc.* **166**, 389–400.

Sletr UB, Messner P, Pum D, Sara M. (1993) Crystalline bacterial cell surface layers: general principles and application potential. *J. Appl. Bacteriol.* **74**, 21S–32S.

Small V, Herzog M, Haner M, Aebi U. (1994) Visualization of actin filaments in keratocyte lamellipodia: negative staining compared with freeze-drying. *J. Struct. Biol.* **113**, 135–141.

Sosinsky G. (1995) Mixing of connexins in gap junction membrane channels. *Proc. Natl Acad. Sci. USA* **92**, 9210–9214.

Stauffer KA, Kumar NM, Gilula NB, Unwin N. (1991) Isolation and purification of gap junction channels. *J. Cell Biol.* **115**, 141–150.

Steven AC, Smith PR, Horne RW. (1978) Capsid fine structure of cowpea chlorotic mottle virus: from a computer analysis of negatively stained virus arrays. *J. Ultrastruct. Res.* **64**, 63–73.

Tranum-Jensen J. (1988) Electron microscopy: assays involving negative staining. In *Methods in Enzymology*, Vol. 165. Academic Press, New York, pp. 357–374.

Tsuprun V, Rajagopal BS, Anderson D. (1995) Electron microscopy of *Bacillus subtilis* GroESL chaperonin and interaction with the bacteriophage ø 29 head–tail connector. *J. Struct. Biol.* **115**, 258–266.

Vonck J, van Bruggen EFJ. (1992) Architecture of peroxisomal alcohol oxidase crystals from the methylotropic yeast *Hansenula polymorpha* as deduced by electron microscopy. *J. Bacteriol.* **174**, 5391–5399.

Walz T, Häner M, Wu X-R, Henn C, Engel A, Sun T-T, Aebi U. (1995) Towards the molecular architecture of the asymmetric unit membrane of the mammalian urinary bladder epithelium: a closed "twisted ribbon" structure. *J. Mol. Biol.* **248**, 887–900.

Ward RJ, Leonard K. (1992) The *Staphylococcus aureus* α-toxin channel complex and the effect of Ca^{2+} ions on its interaction with lipid layers. *J. Struct. Biol.* **109**, 129–141.

Wells B, Horne RW, Shaw PJ. (1981) The formation of two-dimensional arrays of isometric plant viruses in the presence of polyethylene glycol. *Micron* **12**, 37–45.

Welte C, Nickel R, Wild A. (1995) Three-dimensional crystallization of the light-harvesting complex from *Mantioniella squamata* (Prasinophyceae) requires an adequate purification procedure. *Biochim. Biophys. Acta* **1231**, 265–274.

White JG. (1991) The cytoskeleton of human blood platelets. In *Blood Cell Biochemistry* (ed. JR Harris) Vol. 2. Plenum Press, London, pp. 113–148.

Wilmsen HU, Leonard KR, Tichelaar W, Buckley JT, Pattus F. (1992) The aerolysin membrane channel is formed by heptamerization of the monomer. *EMBO J.* **11**, 2457–2463.

Wrigley NG. (1968) The lattice spacing of crystalline catalase as an internal standard of length in electron microscopy. *J. Ultrastruct. Res.* **24**, 454–464.

Wrigley NG. (1979) Electron microscopy of influenza virus. *Br. Med. Bull.* **35**, 35–38.

Wrigley NG, Brown EB, Skehel JJ. (1986) Electron microscopy of influenza virus. In *Electron Microscopy of Proteins* (eds JR Harris, RW Horne) Vol. 5. Academic Press, London, pp. 103–164.

Zahn R, Harris JR, Pfeifer G, Plückthun A, Baumeister W. (1993) Two dimensional crystals of the molecular chaperone GroEL reveal structural plasticity. *J. Mol. Biol.* **229**, 579–584.

6 Unstained Vitrified Specimens: Introduction and Technical Background

Efforts to study unstained, fully hydrated thin film biological specimens in vitreous ice by TEM started some time ago (Fernández-Morán, 1960; Hutchinson et al., 1978; Taylor and Glaeser, 1974, 1976). However, such approaches were not realistically achieved until the development of a method for rapidly freezing thin aqueous films in the early 1980s, by Jacques Dubochet and his colleagues (Dubochet and McDowall, 1981; Dubochet et al., 1982). Biological materials could then routinely be prepared as frozen-hydrated specimens, with untreated material (unfixed and unstained) embedded within a thin layer of vitreous/non-crystalline ice.

Thus, the new field of high resolution biological cryoelectron microscopy using unstained vitrified thin film specimens, produced by the *Adrian method,* was established and rapidly utilized by several research groups around the world for the cryo-TEM study of an increasing number of biological samples. In retrospect, it can now be seen that the considerable time and effort invested in the investigation and understanding of the fundamental technical and physical principles involved in vitrification resulted in its successful application to aqueous suspensions of biological materials (Lepault et al., 1983a, 1985), and helped provide answers to many structural and functional biological questions (for review see Dubochet et al., 1985, 1988; Unwin, 1986). A detailed account of the broad field of cryoelectron microscopy also appears in the book by Echlin (1992) and brief surveys dealing with thin vitrified specimen production are included in the book by Roos and Morgan (1990), and the chapter by Stewart (1991).

For the establishment of the thin film vitreous ice specimen preparation technique (Adrian et al., 1984) and the routine cryoelectron microscopy of these frozen-hydrated specimens the following points were important. It was necessary:

1. To achieve the formation of a thin aqueous film containing a single layer of appropriately spaced biological material that spanned the holes of either a fine mesh bare EM grid or the holes in a holey/perforated carbon support film. Thin aqueous films were obtained by direct but brief filter paper contact (i.e. fluid removal by 'blotting' the excess fluid from specimen grids).

2. To produce specimens containing thin vitreous ice rather than crystalline ice by very rapid freezing of the thin aqueous layer. This was achieved by rapidly plunging a specimen grid, immediately after blotting, vertically into a liquid nitrogen-cooled hydrocarbon cryogen, such as liquid propane or liquid ethane.
3. To develop procedures for the careful handling and storage of the grids containing the vitrified specimens, constantly kept under ice-free liquid nitrogen.
4. To develop systems for the insertion of vitreous ice specimen grids into suitable liquid nitrogen-cooled cryospecimen holders, and for the transfer of such holders to suitably stable stages within the TEM without accumulation of ice on the specimen or a significant amount of ice on the holder, as a result of atmospheric moisture.
5. To develop adequate liquid nitrogen-cooled anti-contaminator devices within the electron microscope, ideally maintained at a lower temperature than that of the specimens, to reduce to a minimum the accumulation of crystalline hexagonal ice particles on the surface of a vitreous ice specimen, or indeed the build-up of vitreous ice (Lepault *et al.*, 1983b) during its time within the high vacuum of the TEM. (In practice, an efficient anti-contaminator, maintaining a temperature of −125°C or lower, should create a very low partial pressure of water within the vacuum system of a TEM.)
6. To study specimens of biological materials in the TEM under low electron dose conditions, with an optimal defocus to provide enhanced phase contrast within the image, in order to retrieve detailed structural information otherwise not available from the extremely low amplitude contrast of the unstained images.

Some of the low temperature technical and electron microscopical aspects included under (4), (5) and (6) above, were established during earlier electron crystallographic studies on thin enzyme crystals such as catalase (Chiu, 1982; Taylor and Glaeser, 1974, 1976). These studies provided a valuable source of important technical information for low temperature studies of thin specimens and enabled these and other groups immediately to be in a position to investigate the possibilities of cryoelectron microscopy of thin unstained vitrified specimens of biological materials.

The technically important room temperature low electron dose crystallographic studies of Unwin and Henderson (1975), on unstained glucose-embedded catalase and purple membrane were subsequently extended to low temperatures (Henderson *et al.*, 1986, 1990; Milligan *et al.*, 1984; Unwin and Ennis, 1984). These workers and others have continued to successfully pursue the study of thin crystalline samples at low temperatures, in the presence of glucose, trehalose or tannic acid (Chiu, 1982, 1986, 1993; Jap and Li, 1995; Jap *et al.*, 1990, 1991; Kühlbrandt, 1992; Nogales *et al.*, 1995) and in the presence of vitreous ice alone (Asturias and Kornberg, 1995; Brisson *et al.*, 1994).

From an initially small number of electron microscopists pursuing the thin film–vitreous ice approach, the number of functional cryoelectron microscopes throughout the world has steadily increased since the mid-1980s, with a subsequent increase in the volume of published data. Foremost in this expansion, have been investigators such as Marc Adrian, Tim Baker, Steve Fuller, Elizabeth Hewat, John Kenney, Peter Metcalfe and Alasdair Steven (viral structure), Nigel Unwin, Carmen Manella and Christopher Akey (membranes), Peter Frederick, David Siegel and Yshi Talman (surfactants and lipids), Felix de Haas, Marin van Heel, Joachim Frank, Nicholas Boisset, Michael Radermacher and Helen Saibil (single molecule/particle analysis), Jean-Francois Menetret, Rasmus Schröeder and John Trinick (muscle proteins), Dick Wade, Eckhardt and Eva Mandelkow and Ken Downing (microtubule structure) and David DeRosier (flagellar structure). Jacques Dubochet continues to play a leading role in cryoelectron microscopy, particularly within the field of nucleic acids and nucleoprotein complexes. Others have been more particularly concerned with further technical development of the rapid freezing process (Battersby *et al.*, 1994; Trachtenberg, 1993), an aspect of vital importance for the extension of the technique for the investigation of rapid dynamic events. Examples from the cryoelectron microscopical studies of some of the aforementioned scientists will be given in Chapters 8, 9 and 10.

References

Adrian M, Dubochet J, Lepault J, McDowall AW. (1984) Cryo-electron microscopy of viruses. *Nature* **308**, 32–36.

Asturias FJ, Kornberg RD. (1995) A novel method for transfer of two-dimensional crystals from the air/water interface to specimen grids. *J. Struct. Biol.* **114**, 60–66.

Battersby BJ, Sharp JCW, Webb RI, Barnes GT. (1994) Vitrification of aqueous suspensions from a controlled environment for electron microscopy: an improved plunge-cooling device. *J. Microsc.* **176**, 110–120.

Brisson A, Olofsson A, Ringler P, Schmulz M, Stoylova S. (1994) Two-dimensional crystallization of proteins on planar lipid films and structure determination by electron microscopy. *Biol. Cell* **80**, 221–228.

Chiu W. (1982) High resolution electron microscopy of unstained hydrated protein crystals. In *Electron Microscopy of Proteins* (ed. JR Harris) Vol. 2. Academic Press, London, pp. 233–259.

Chiu W. (1986) Electron microscopy of frozen, hydrated biological specimens. *Ann. Rev. Biophys. Chem.* **15**, 237–257.

Chiu W. (1993) What does electron cryomicroscopy provide that X-ray crystallography and NMR spectroscopy cannot? *Ann. Rev. Biophys. Biomol. Struct.* **22**, 233–255.

Dubochet J, McDowall AW. (1981) Vitrification of pure water for electron microscopy. *J. Microsc.* **124**, RP3–4.

Dubochet J, Chang J-J, Freeman R, Lepault J, McDowall AW. (1982) Frozen aqueous suspensions. *Ultramicroscopy* **10**, 55–62.

Dubochet J, Adrian M, Lepault J, McDowall AW. (1985) Cryo-electron microscopy of vitrified biological specimens. *Trends Biol. Sci.* **10**, 143–146.

Dubochet J, Adrian M, Chang J-J, Homo J-C, Lepault J, McDowall AW, Schultz P. (1988) Cryo-electron microscopy of vitrified specimens. *Quart. Rev. Biophys.* **21**, 129–228.

Echlin P. (1992) *Low Temperature Microscopy and Analysis.* Plenum Publishing Corporation, New York.

Fernández-Morán H. (1960) Low temperature preparation techniques for electron microscopy of biological specimens based on rapid freezing with liquid helium. II. *Ann. NY Acad. Sci.* **56**, 801–808.

Henderson R, Baldwin JM, Downing KH, Lepault J, Zemlin F. (1986) Structure of purple membrane from *Halobacterium halobium*: recording, measurement and evaluation of electron micrographs at 3.5 Å resolution. *Ultramicroscopy* **19**, 147–178.

Henderson R, Baldwin JM, Ceska TA, Beckmann E, Zemlin F, Downing K. (1990) A model for the structure of bacteriorhodopsin based on high resolution electron cryomicroscopy. *J. Mol. Biol.* **213**, 899–929.

Hutchinson TE, Johnson DE, Mackenzie AP. (1978) Instrumentation for direct observation of frozen hydrated specimens in the electron microscope. *Ultramicroscopy* **3**, 315–324.

Jap BK, Li H. (1995) Structure of the osmo-regulated H_2O-channel AQP-CHIP, in projection at 3.5 Å resolution. *J. Mol. Biol.* **251**, 413–420.

Jap BK, Downing KH, Walian PJ. (1990) Structure of PhoE porin in projection at 3.5 Å resolution. *J. Struct. Biol.* **103**, 57–63.

Jap BK, Walian PJ, Gehring K. (1991) Structural architecture of an outer membrane channel as determined by electron crystallography. *Nature* **350**, 167–170.

Kühlbrandt W. (1992) Two-dimensional crystallization of membrane proteins. *Quart. Rev. Biophys.* **25**, 1–49.

Lepault J, Booy FP, Dubochet J. (1983a) Electron microscopy of frozen biological suspensions. *J. Microsc.* **129**, 89–102.

Lepault J, Freeman R, Dubochet J. (1983b) Electron beam induced 'vitrified ice'. *J. Microsc.* **132**, RP3–4.

Lepault J, Pattus F, Martin N. (1985) Cryo-electron microscopy of artificial biological membranes. *Biochim. Biophys. Acta* **820**, 315–318.

Milligan RA, Brisson A, Unwin PNT. (1984) Molecular structure of crystalline specimens in frozen aqueous solutions. *Ultramicroscopy* **13**, 1–10.

Nogales E, Wolf SG, Zhang SX, Downing KH. (1995) Preservation of 2-D crystals of tubulin for electron crystallography. *J. Struct. Biol.* **115**, 199–208.

Roos N, Morgan AJ. (1990) Imaging and analysis of frozen-hydrated specimens in the TEM: vitrified thin films and cryosections. In *Cryopreparation of Thin Biological Specimens for Electron Microscopy: Methods and Applications* (RMS Handbook no. 21). Oxford University Press, Oxford, pp. 50–68.

Stewart M. (1991) Transmission electron microscopy of vitrified biological macromolecular assemblies. In *Electron Microscopy in Biology: a Practical Approach* (ed. JR Harris). IRL Press, Oxford, pp. 229–242.

Taylor KA, Glaeser RM. (1974) Electron diffraction of frozen, hydrated protein crystals. *Science* **186**, 1036–1037.

Taylor KA, Glaeser RM. (1976) Electron microscopy of frozen hydrated biological specimens *J. Ultrastruct. Res.* **55**, 448–456.

Trachtenberg S. (1993) Fast-freezing devices for cryo-electron-microscopy. *Micron* **24**, 1–12.

Unwin PNT. (1986) The use of cryoelectron microscopy in elucidating molecular design and mechanisms. *Ann. NY Acad. Sci.* **483**, 1–4.

Unwin PNT, Ennis PD. (1984) Two configurations of a channel-forming membrane protein. *Nature* **307**, 609–612.

Unwin PNT, Henderson R. (1975) Molecular structure determination by electron microscopy of unstained crystalline specimens. *J. Mol. Biol.* **94**, 425–440.

7 Unstained Vitrified Specimens: Preparation Procedures

7.1 Routine plunge freezing of biological samples

7.1.1 Vitrification on bare grids or holey carbon films

This technique can be performed using equipment of the type shown in *Figures. 7.1* and *7.2*, as described in the legends to these figures. Similar equipment is available commercially, for example the Reichert KF80 (sometimes termed the 'cryostation', plunge freezer or immersion cryofixer). Equipment of the type shown in *Figure 7.1* can be constructed by skilled personnel in a reasonably well-equipped mechanical workshop (technical details may be obtained from the EMBL, Heidelberg). A variety of smaller routine accessories are also needed for working under liquid nitrogen, such as forceps, small screwdriver, metal probes and small plastic or metal grid containers or boxes. Individual laboratories rapidly establish their own working system for cryomicroscopy, centred around the main procedure described below. As with many other technical procedures, the simplest approach often proves to be the most successful.

For vitrification, an aqueous suspension of biological material between 1–2 mg ml^{-1} (protein, nucleic acid or lipid) will usually be adequate. Lower concentrations generally lead to excessive spacing of the biological particles within the thin layer of vitreous ice and higher concentrations can lead to overlapping of particles. In practice, however, the usable concentration range of the sample may be rather broad (10 µg ml^{-1} to 5 mg ml^{-1}). The presence of isotonic salt and low concentrations of buffer solutions does not create any difficulties, although samples that have been dialysed against distilled water are most satisfactory, since this generates the lowest possible density vitreous ice layer to embed the slightly higher density biological material. In the absence of excessive evaporation before freezing, the specimen contrast change up to 10 mM salts is negligible. The advantage of pure water is that it tolerates evaporation without any osmotic effect: 99% evaporation from 10^{-4} M gives 10^{-2} M, which is still acceptably low. The disadvantage of almost pure water is the large Debye

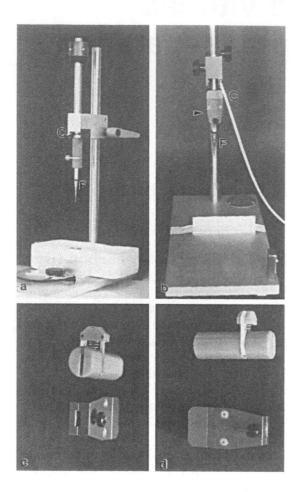

Figure 7.1: Gravity plunge freezing apparatus.
Gravity plunge freezing apparatus for the production of thin vitrified specimens. (a) The standard apparatus with spring-type forceps holder, and (b) a newer apparatus with clamp-type forceps holder and an in-built magnet to hold the forceps when the clamp is loose. (c, d) Close-up views of the two types of forceps holder [(F) forceps (C) cable release mechanism for plunger].

length of approximately 300 at 10^{-4} M which generates long-distance repulsion.

Experience of freezing samples that contain high concentrations of reagents such as urea or guanidine hydrochloride is limited. These substances may cause specimen bubbling in the electron beam and should be avoided, as should solutions containing sucrose and glycerol at a concentration above a few percent. However, recent observations (Cyrklaff *et al.*, 1994) have shown that in thin films protein, DNA molecules and viral particles are 'held' at the fluid–air interface. Repeated washing/microdialysis of the interface sample can then be performed directly on the grid with approximately 10 µl droplets of buffer or water, immediately before freezing. The orientation of particles within a thin aqueous layer may be considerably influenced by surface tension/interfacial forces and particles may move to regions of thicker aqueous film at the edge of the grid squares or holes in the carbon film. Also, in regions of very thin ice, particles may even protrude slightly from the surface of the vitreous ice (Cyrklaff *et al.*, 1994) particularly if there is any significant sample freeze-drying within the vacuum of the electron microscope. As the specimen grid is usually maintained at a temperature of −170°C or lower, freeze-drying should be negligible.

Procedure. For safety reasons, the procedure to be described should be performed within a fume extraction hood/cabinet. Thin aqueous films can be produced by applying approximately 5 µl of biological sample to a bare 400- to 600-mesh EM grid, followed by the removal of excess fluid by blotting on to a filter paper (e.g. Whatman No. 1) for 1–3 sec. This leaves a very thin layer of fluid spanning the grid holes. The grid should then be instantly plunge frozen (within approx. 0.5 sec) in liquid ethane or propane, before significant evaporation occurs from the thinly spread sample. The vitreous ice layer towards the centre of the grid squares will rou-

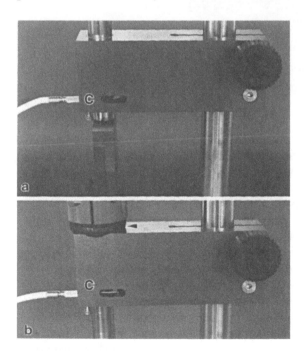

Figure 7.2: Release mechanism of the rapid freezing apparatus.
Close-up photographs showing the release mechanism of the rapid freezing apparatus in *Figure 7.1b*, (a) before, and (b) after pressing the cable release and rapid free-fall of the mechanism on to its rubber stop (arrow).

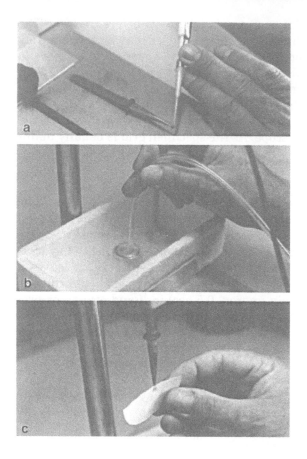

tinely be found to be slightly thinner than that closer to the grid bars.

The alternative, and perhaps more widely used approach, is to use holey carbon support films that have been glow-discharge treated to assist fluid spreading. Holey carbon support films that have been further stabilized by the deposition of a fine layer of gold by sputter-coating (primarily as a focusing aid; see *Figure 3.7*) can also be used; in this instance glow-discharge treatment is not usually necessary. The biological sample should again be applied as a 3–5 μl droplet to the holey carbon surface, excess fluid removed by blotting and the grid immediately plunge frozen.

Liquid ethane (at approx. −160°C) should be prepared in advance within a small container surrounded by liquid nitrogen, by condensing ethane gas and waiting until the surface of the ethane appears slightly cloudy. Alternatively, frozen ethane can be partly thawed by the insertion of a metal probe. Liquid ethane has a higher vapour pressure than propane, and residual frozen ethane is more readily removed from the specimen grid within the vacuum of the cryoelectron microscope. For laboratory safety reasons, it is better to use cylinders of ethane rather than propane. Grids should be held in very lightly silicone-treated forceps clamped in

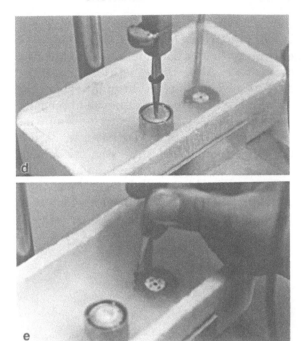

◀ ▲   **Figure 7.3:**   Production of a thin film vitrified specimen.
The overall sequence of operations during the production of a thin film vitrified specimen; see text for details. (a) Application of a sample to holey carbon film, (b) filling of liquid ethane container, within liquid nitrogen bath. (c) Blotting of specimen grid, (d) after rapid plunge of the specimen grid into liquid ethane, and (e) removal of forceps and grid from the plunge freezer and storage of the vitrified specimen. All photographs courtesy of Marc Adrian.

the plunge freezing apparatus and sample blotting should be performed *immediately* before freezing. A small rubber O-ring or freely sliding ring of plastic tubing can be used to hold the forceps closed, or alternatively reverse action forceps can be used. A photographic release cable or an electrical foot-switch is used to rapidly release the gravity-assisted free-fall plunge freezing mechanism, adjusted so that the forceps + grid speedily enter the container of liquid ethane surrounded by liquid nitrogen. The forceps + vitrified specimen should then be released from the apparatus, fluid cryogen instantly removed and the grid transferred to an adjacent container submerged in liquid nitrogen. Occasionally an additional pair of forceps may be required to release the grid from the first pair of forceps (it can be held by a small amount of frozen hydrocarbon cryogen). Small aluminium bottle-caps or purpose-built plastic grid holders are most suitable for this initial accumulation of vitrified specimens, under liquid nitrogen. The overall sequence for the production of thin frozen-hydrated specimens is depicted in *Figure 7.3*.

Apparatus has been designed and constructed (Dr Marc Adrian, Laboratory of Ultrastructural Analysis, University of Lausanne) in an attempt to standardize and partially automate the sample blotting time and rapid freezing procedure (*Figure 7.4*). In practice such additional equipment is usually only of marginal benefit, since vitrified specimens are produced in rather small quantities at any one time. Also, with repetition, any scientist or technician will become progressively skilled at the manual blotting and rapid freezing procedure. Evidence that the thin aqueous film spans one surface of the holey carbon film immediately prior to freezing (the surface that received the glow-discharge treatment) is shown in *Figure 7.5*. A frozen hydrated thin film of 5% trehalose containing bacteriophage particles was imaged in the cryo-TEM (*Figure 7.5a*) and then allowed to freeze-dry. The thin trehalose film remained intact and the grid was then imaged by scanning electron microscopy (*Figure 7.5b*), from which it is evident that the thin film spans only one edge of the holes (Marc Adrian, unpublished observations).

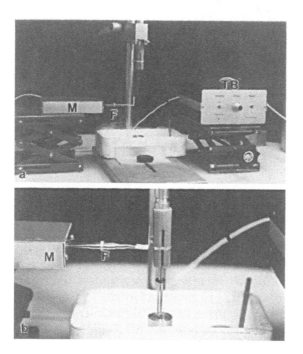

Figure 7.4: Semi-automatic specimen blotting and timing device.

(a) A semi-automatic specimen blotting and timing device linked to the plunge freezing apparatus (produced by Marc Adrian). A piece of right-angle bent filter paper is held by a pair of forceps (F), themselves held by a magnetically controlled positioning device (M). The paper and the positioning device are positioned a small distance away from the sample droplet on the grid, such that when the device is activated the forceps move the paper just into contact with the droplet on the grid. The sequence of events is activated and controlled by triggering the timing box (TB). After the filter paper touches the grid, a period of seconds (1–3 sec) is allowed for fluid removal. The forceps and filter paper are then automatically withdrawn, followed instantly by the triggering of the release mechanism of the plunge freezing apparatus (b).

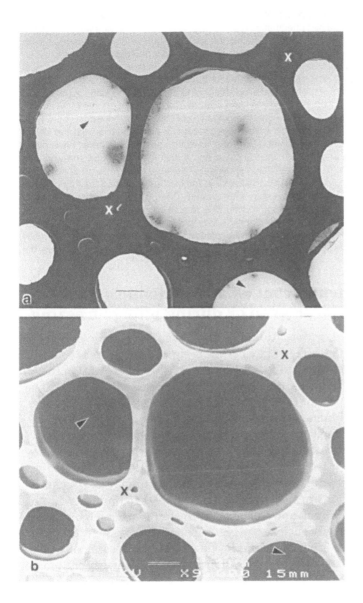

Figure 7.5: A vitrified specimen of bacteriophage ϕ29 imaged by cryoelectron microscopy and SEM.

Low magnification field of a vitrified specimen of bacteriophage ϕ29 prepared from 5% trehalose. (a) Imaged by cryoelectron microscopy. (b) The *same region of the specimen* imaged by SEM following freeze-drying of the thin vitreous ice film. Note the reference points x and the fact that the SEM specimen has been slightly tilted. The thin dried film of trehalose (b) is clearly shown to span the futhermost edge of the holey carbon. Breaks are present in both the vitreous and freeze-dried film (arrows). At even lower magnification, thickening of the freeze-dried film at the edges of the holes and pronounced undulation of the holey carbon support film across the grid, particularly near the grid bars, are also apparent from SEM images. The scale bars indicate 1 µm. Micrographs courtesy of Marc Adrian.

▶ **Figure 7.6:** A self-contained pneumatically operated rapid freezing device.

An example of a self-contained pneumatically operated rapid freezing device with retractable environmental facility (courtesy of Shlomo Trachtenberg). An overview of the apparatus in its starting position is shown (full technical information will be pubished elsewhere by Dr Trachtenberg). Access to the specimen grid inside the Perspex environmental chamber (arrows) is through a side blotting port, which together with a microprocessor and thermocouple-controlled heating element at the bottom of the chamber enables considerable versatility of the experimental design and control of the thin aqueous sample (temperature and humidity), prior to rapid plunge freezing into the liquid ethane container within the polystyrene liquid nitrogen vessel (V). C1, air cylinder driving chamber; C2, air cylinder driving specimen; C3, air cylinder driving the Teflon shutter between the environmental chamber and the freezing vessel; LBP, lower base plate; UBP, upper base plate; P, pneumatics, valves, gates and manifolds to hook-up spraying and illumination device and control individual piston speeds; R, stainless steel rods to support base plates and guide the Perspex chamber, which slides on Teflon bushes. S1, switch to automatically operate the system in the following sequence: open shutter (C3); plunge specimen (C2); raise chamber (C1). S2, switch to reverse S1 in the following sequence: lower chamber (C1); raise specimen (C2); close shutter (C3). TB, teflon bushes; T1, toggle switch to release the pressure from C1 for manual handling; T2, toggle switch to release the pressure from C2 to manually drive the specimen; T3, toggle switch to release the pressure from C3 to manually drive the shutter.

7.1.2 Environmental control and dynamic experiments prior to or during plunge freezing

As already mentioned, evaporation of water from the sample can occur after blotting and before freezing, causing a significant increase in salt concentration and generating an undesirable structural alteration to the biological material (Trinick and Cooper, 1990; Walker *et al.*, 1994). The enclosure of the top part of the cryostation containing the blotted specimen grid within a cabinet, to maintain high humidity, can reduce evaporation from the specimen to a minimum. This arrangement also allows the variation and control of sample temperature prior to freezing (Battersby *et al.*, 1994; Bellare *et al.*, 1986, 1988; Frederick *et al.*, 1991). A simple first approach to environmental control is to position the plunge freezing apparatus within a cooled or warmed room, within which the humidity can be increased if desired. Trinick and Cooper (1990) simply used a moist filter paper 'sandwich' to briefly enclose the blotted specimen grid, and thereby prevent sample evaporation. An example of a versatile environmental system (not commercially available) is shown in *Figure 7.6* (courtesy of Shlomo Trachtenberg).

Further extension of this approach enables dynamic biochemical, physiological and pharmacological treatments to be applied to the sample within the thin aqueous film immediately before (Walker *et al.*, 1994) or after blotting, with very short interaction times (i.e. during the very rapid fall, immediately prior to freezing). For time-resolved processes, such as the rapid opening of the acetylcholine receptor channel without reclosing due to the refractory state, this approach has proved to be extremely useful (Unwin, 1995; see also Chapter 10). The application of transmitters, substrates, drugs and pH alteration by spraying microdroplets on to the sample grid has been successful in the investigation of such rapid events

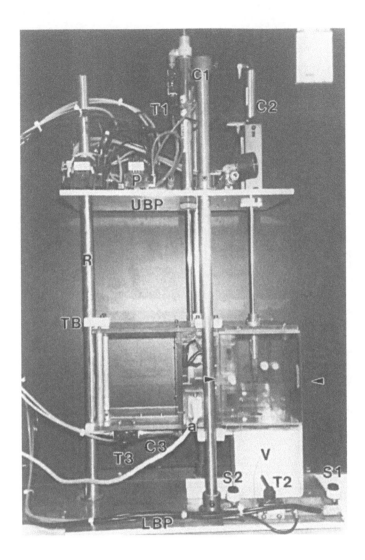

(see *Figure 7.7* from Berriman and Unwin, 1994). Incubation of biological samples at controlled temperatures prior to the production of the thin aqueous films may be adequate for the investigation of slower dynamic events, such as lipid phase changes.

In addition, photolabile substances and rapid light-induced structural changes can be investigated by briefly illuminating the falling specimen grid with a laser beam (Subramanian *et al.*, 1993). Similarly, a light beam can be used to produce an extremely rapid and controllable increase in temperature within the sample in the thin fluid layer by illumination from a flash tube (temperature jump) for a known length of time whilst the falling grid passes through the light beam, immediately prior to freezing (Siegel *et al.*, 1994).

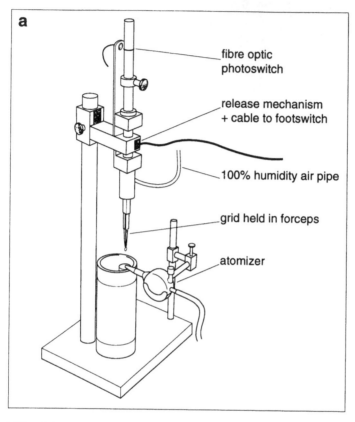

fibre optic
photoswitch

release mechanism
+ cable to footswitch

100% humidity air pipe

grid held in forceps

atomizer

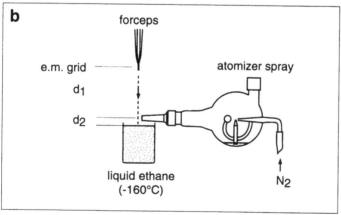

forceps

e.m. grid

d_1

d_2

atomizer spray

liquid ethane
(-160°C)

N_2

◀ **Figure 7.7:** Spray freezing apparatus.
A diagrammatic example of a spray freezing apparatus (a) consisting of a guillotine-type plunger and an atomizer spray. The plunger is released by a footswitch, and the movement activates a fibre optic photoswitch, which in turn triggers the opening of a solenoid valve for nitrogen gas (1.6 bar) to set off the spray (e.g. containing acetylcholine, ATP and other enzyme substrates, drugs, cations, etc.). To prevent frosting of the spray nozzle the liquid-nitrogen-cooled ethane and Dewar are moved into place just before spraying. The whole apparatus is enclosed within a high-humidity chamber. (b) Details of the atomizer spray. The electron microscope grid is held in a pair of forceps at a fixed distance (d_1) above the nozzle of the atomizer, which is at a fixed distance (d_2) above the surface of the liquid ethane. After applying the specimen to the grid and blotting off the excess solution, it is allowed to rapidly free-fall into the liquid ethane, briefly passing through the spray. Diagrams courtesy of Nigel Unwin. Reprinted from Berriman and Unwin (1994) Analysis of transient structures by cryo-microscopy combined with rapid mixing of spray droplets. *Ultramicroscopy* 56, 241–252, with permission from Elsevier Science-NL.

7.2 Storage of vitrified specimens

Once produced, vitrified specimens can be kept for long periods of time under dry liquid nitrogen. For convenience, grids can be transferred to small circular grid boxes, which are available commercially or can be prepared in a workshop from the larger plastic sliding-top boxes. These small storage containers usually have a rotating top and enable single grids to be positioned in four places (*Figure. 7.8*). The manipulation of the tops of these boxes, for grid insertion and removal whilst under liquid nitrogen is not easy. Nevertheless, with a little practice, this usually proves possible and apart from long-term storage the boxes are useful for transporting specimen grids, under liquid nitrogen in a small Dewar flask. The circular grid boxes can be numbered and stored in larger Dewar flasks, in holders similar to those employed for the long-term storage of cultured cells (*Figure 7.8*).

7.3 Cryotransfer of vitrified specimens to the electron microscope

The successful insertion of a specimen grid containing vitrified material into a liquid nitrogen-cooled EM specimen holder, with rapid transfer of the holder into a suitable cryo-TEM without accumulation of crystalline ice on the specimen, was initially achieved by several scientists using in-house modified specimen holders (e.g. for the Philips TEMs) and subsequently suitable holders and additional equipment became available from manufacturers, such as Oxford Instruments and Gatan (*Figure 7.9*). These are available in various modifications to fit most commercial TEMs suitable for cryostudies and technical improvement continues. The clean high vacuum of the cryo-TEM (10^{-6} Torr) needs to have an extremely low water

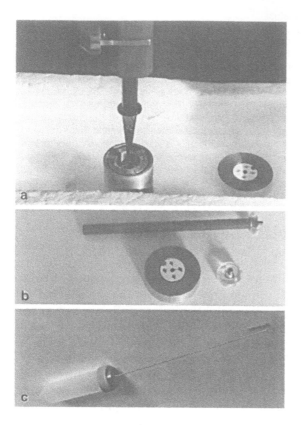

Figure 7.8: Plastic holder for the storage of vitrified specimens.
An example of a small plastic holder for the low temperature storage of vitrified specimens. (a) Supported in cooled metal holder under liquid nitrogen for short-term storage. (b) Transport rod for plastic grid holder and holder lid. (c) Container for long-term storage of a number of grid-holders under liquid nitrogen (e.g. within a cell-storage Dewar flask). Although available commercially, holders of this type can readily be constructed by a technician in a mechanical workshop.

vapour content (partial pressure less than 5×10^{-8} Torr); this is achieved by a liquid nitrogen pre-cooled anticontaminator system which prevents vitreous or crystalline ice particles forming on the specimen. A sliding shield at the end of the specimen rod protects the specimen throughout the transfer procedure (*Figure 7.10*) and is not removed until temperature stabilization of the specimen holder within the TEM has been achieved (usually monitored at approx. −160 to −180°C). If possible, the temperature of the principal anti-contaminator should also be monitored and maintained at a lower temperature than that of the specimen. Efficient anticontaminators are of the flat blade/fork-type, and provide cooled surfaces close to and at both sides of the specimen grid (*Figure 7.11*). They must be carefully centred in advance (with respect to the electron beam and height) otherwise the tilting of the specimen rod may be restricted.

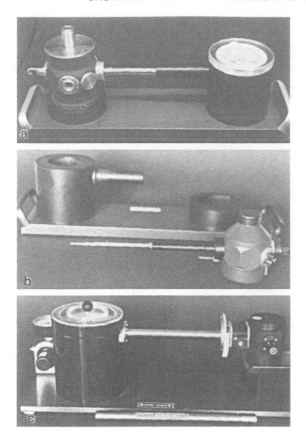

Figure 7.9: Commercially available cryospecimen holders.
Examples of commercially available cryospecimen holders (a, b) from Gatan and (c) from Oxford Instruments. Each cryoholder comes with its own miniature Dewar flask and workstation.

The temperature of the specimen must always be maintained lower than −125°C in order to preserve the vitreous water; above this temperature, vitrified water is metastable and readily transforms to crystalline cubic ice. Vitrified water can be identified in the TEM by an electron diffraction pattern showing fuzzy broad diffraction rings, compared to the distinct rings of crystalline cubic ice. If the initial vitrification procedure has been performed in a satisfactory manner, the thickness of the vitreous water/ice may actually increase during time in the electron microscope.

Despite the above precautions, over a period of hours microcrystals of hexagonal ice may accumulate upon the vitrified specimen, a progressive alteration that appears to be potentiated by the presence of salts within the thin vitrified layer; salt may provide nucleation sites for ice crystal growth. This specimen deterioration appears to vary with vacuum sys-

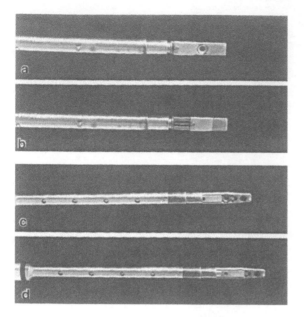

Figure 7.10: Cryospecimen holder rods.
Examples of rods of cryospecimen holders from Gatan (a,b) and Oxford Instruments (c, d). The sliding shield that totally encloses the specimen grid is open (a and c) and closed (b and d).

tems of cryoelectron microscopes from different manufacturers; there is apparently no such problem with the most recent instruments.

The precise sequence of events for the cryotransfer will be determined by the equipment available and most individuals will require considerable practice before the sequence of operations can be performed rapidly and smoothly without damage to the specimen.

The following description is of a generalized procedure. A single grid containing a vitrified specimen should be transferred to the tip of an appropriate cryoholder under liquid nitrogen, located within a purpose-designed cryotransfer workstation (*Figure 7.12*). The cryoholder comes with its own miniature Dewar flask, has a temperature monitoring system and usually an integral heater, to increase the temperature of the holder tip and specimen grid, if this is required. Initially, equipment should always be used in the manner described by the manufacturer, before minor innovations and improvements are introduced. The tip of the cryoholder is sometimes stored under vacuum prior to cooling, to decontaminate and minimize the presence of metal-bound water in the vicinity of the specimen grid (e.g. the Oxford Instruments cryoholder).

After insertion into the work-station the chamber is filled with liquid nitrogen (*Figure 7.13a*), but the miniature Dewar flask of the specimen holder need not be filled at this stage. Whilst under liquid nitrogen, the sliding shield at the end of the holder should be removed (*Figure 7.13b*) and the specimen grid positioned (*Figure 7.14a*) and held firmly, usually by a screw insert to ensure good mechanical and thermal contact. The specimen grid should then be 'washed' with dry liquid nitrogen (*Figure 7.14b*). The sliding shield should then be returned, to protect the grid from atmospheric water vapour during the rapid transfer of the cryoholder to the TEM.

Early cryoholders used clips to hold the grid in position; this was often unsatisfactory because of interference by frozen organic cryogen. A steady improvement in the design and performance of liquid nitrogen conductivity-cooled cryoholders has been seen over the past 10 years, since the original circulating nitrogen gas-cooled specimen holder was produced by the Philips company (see data sheets from Gatan and Oxford Instruments).

The cryoworkstation, with cryoholder still in position, should then be positioned on the flat surface of the TEM next to the column, the cryoholder removed and rapidly inserted into the specimen airlock. The pumping system of the specimen airlock will be activated instantly (*Figure 7.15a*).

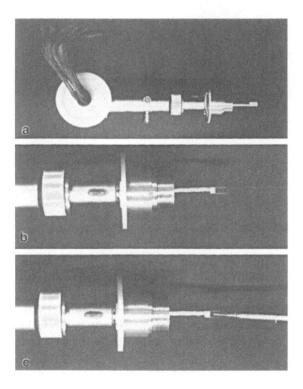

Figure 7.11: Fork-type double blade anti-contaminator.
An example of an efficient fork-type double blade anti-contaminator, as for the Philips TEMs. (a) The whole device, with copper cooling strands, x/y and height controls. (b) A side view of the two flat blades of the anti-contaminator, and (c) with the tip of a specimen rod inserted between the blades. Such a device is now a standard attachment on most cryo-TEMs. Equipment and photographs courtesy of Marc Adrian.

Figure 7.12: Cryospecimen holder.
A tip of a cryospecimen holder inserted within a cryotransfer workstation; (a) Gatan, (b) Oxford Instruments.

Sometimes a double evacuation cycle of the specimen airlock can be performed to reduce the water vapour to a minimum. The specimen holder should then be inserted into the high vacuum (*Figure 7.15b*) and the miniature Dewar flask filled with liquid nitrogen. The temperature of the tip of the cryoholder can be monitored throughout the transfer (it should not exceed −160°C) and subsequent temperature stabilization to the level maintained by the liquid nitrogen in the cryoholder Dewar (*Figure 7.16*). TEM study can commence after removing the sliding shield from around the specimen.

The cryoholder of Oxford Instruments has a retractable bellows system surrounding the rod of the holder, which is continually purged by a flow of dry nitrogen gas. This reduces the condensation of atmospheric water vapour on to the cool rod during the rapid transfer of the holder from the workstation to the specimen airlock and into the high vacuum of the TEM. A flow of dry nitrogen can also be supplied to the specimen airlock, prior to insertion of the specimen rod. The combination of these

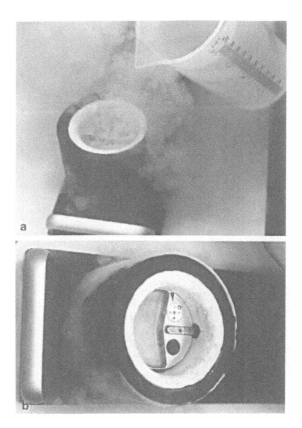

Figure 7.13: Filling the specimen chamber with liquid nitrogen.
With the cryospecimen holder inserted in the workstation and the sliding shield withdrawn, the specimen chamber is filled with liquid nitrogen (a). (At this stage the minature Dewar flask at the end of the holder can be filled, but because of the rotation of the holder during specimen insertion to the air-lock of the TEM, filling is usually left until later; see *Figure 7.15.*) The vitrified specimen grids are then positioned in the holder compartment (arrow, b). The necessary liquid nitrogen pre-cooled manipulation tools (forceps, probes, screw insert and holder, etc.) must be at hand.

two features reduces the time required for decontamination of the holder within the TEM and also enables a succession of cryotransfers to be carried out without interruption, due to minimal ice accumulation on the holder tip and rod. However, some workers think that such additional technical complications are of little benefit and do not assist the overall speed and simplicity of the cryotransfer procedure.

With experience, the complete specimen vitrification and cryotransfer procedure can usually be performed in less than 30 min, but further time will be needed for high vacuum improvement and temperature equilibration before the sliding shield over the specimen grid can be removed and the electron beam turned on. Regular attention must be paid to the liquid nitrogen levels within the anticontaminator Dewars and the miniature

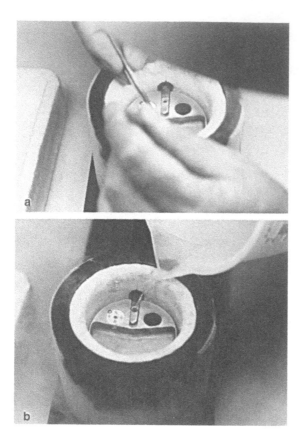

Figure 7.14: Insertion of a vitrified specimen into a cryoholder.
(a) The positioning of a vitrified specimen grid *under liquid nitrogen* into a cryoholder, followed by insertion of the holding ring and washing with ice-free liquid nitrogen (b) before closing the sliding shield and rapid transfer to the cryo-TEM specimen airlock.

Dewar of the specimen holder. Overfilling of the latter should be avoided, since this can cause boiling of the liquid nitrogen and mechanical instability. Simple procedures for checking the liquid nitrogen level and for siphoning-off small amounts of liquid nitrogen from the cryoholder Dewar have therefore been devised.

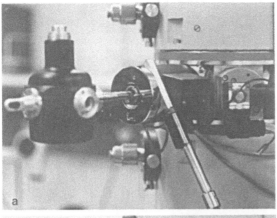

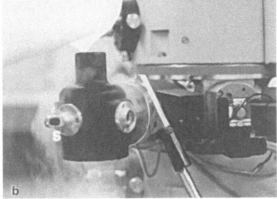

Figure 7.15: Insertion of a cryospecimen holder into a TEM.
Insertion of the cryospecimen holder into the specimen airlock (a) and then into the high vacuum of the electron microscope, filling the small Dewar flask at the end of the cryoholder with liquid nitrogen *before* any significant warming of the specimen occurs (b). [(S) external control for sliding shield.]

7.4 Cryoelectron microscopy of thin vitrified specimens: instrumental considerations

Initially the low electron dose system for the particular cryo-TEM being used needs to be correctly adjusted, with respect to the search mode, focus mode and photography mode. Electron micrographs should always be recorded from regions that have been minimally irradiated at low magnification and then irradiated only during photography when at higher magnification. The defocus in the beam deflected focus mode must be manually calibrated with respect to the undeflected beam in the photographic

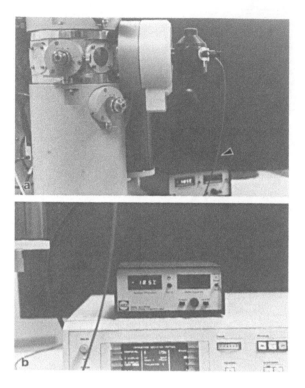

Figure 7.16: Temperature monitoring of the cryospecimen holder.
Temperature monitoring of the cryospecimen holder with a peripheral cold stage control unit (Gatan). (a) The thermocouple connection from the cryoholder to the nearby control unit (arrow). When specimen temperature stability has been reached, the sliding shield around the grid can be removed and TEM study can commence (b).

mode, at any given instrumental magnification, and astigmatism in both modes should be adjusted in advance. This will be described in the manufacturer's instructions and with the introduction of automated and computer controlled TEMs, the storage of the various instrumental settings required by different users has now become routine.

The procedural outline given below is a generalization. The most effective way of learning is to visit a laboratory (or manufacturer's applications department) which uses identical equipment to one's own. In addition, there are a number of useful courses on cryoelectron microscopy or TEM and image processing workshops, organized by the European Molecular Biology Laboratory (Heidelberg), Philips (Eindhoven) and groups in the USA (e.g. National Center for Macromolecular Imaging, Baylor College of Medicine, Houston).

Procedure. Vitrified specimens should always be studied initially in the instrumental *search mode* at low magnifications (less than × 8000, if possible) with minimum illumination of areas of possible interest spanning the grid squares or holes in the carbon support film; these should be rapidly identified and centred. For convenience, the location of a number of such areas can be stored within the computer of most modern TEMs, for subsequent rapid return and further study. Then, at high magnification (e.g.× 200 000) in the *focus mode*, with the beam deflected away from the region to be photographed eventually, the image should be taken to true focus and a pre-calibrated and pre-selected level of defocus produced by a known number of clicks of the fine focus control. Normally, a defocus in the order of 500–5000 nm will be required to provide the necessary phase-contrast enhancement of image features. Images close to instrumental (Scherzer) focus will be found to possess very little contrast, due to the presence of amplitude contrast alone. It is difficult if not impossible to extract structural information from such images, because of the weak phase-contrast transfer and the low amplitude contrast (Toyoshima and Unwin, 1988). In the *photographic/exposure mode*, beam blanking above the specimen comes into operation whilst the photographic plate moves into place, so that the region to be illuminated receives electrons only during the precalibrated exposure, routinely of 1 or 2 sec.

Following photography it is possible with relatively large specimen particles (e.g. many viruses and molluscan hemocyanin) to check by direct observation under conventional higher electron dose the area just photographed, unless a repeat photographic recording at a different defocus is required or even a through focus or tilt series. Thus, sometimes an assessment of the information on the photograph can be made, although this will rapidly damage the specimen. In most cases, however, where the material under investigation is rather small (e.g. small viruses, protein molecules of mass less than 1 MDa and DNA) the images are photographed 'blind', without any knowledge of the content of the electron micrograph until the film has been developed. On returning to the low magnification search mode, the regions irradiated during focusing and photography can be seen and a new area of interest (not previously irradiated) can be centred for study and repetition of the focus and photography cycle.

Low dose electron micrographs, routinely taken on Kodak S0-163 film or Agfa Scientia film, need to be developed in full strength developer (e.g. Kodak D19) for a lengthy period such as 12 min, with constant agitation. Prior calibration of the low dose exposure for the production of a satisfactory emulsion density (e.g. OD 1.0) during development is necessary. The actual electron dose given during image exposure, together with that during the search procedure can be measured by modern TEMs, and total values usually less than 10 e$Å^{-2}$ can be achieved. Indeed, for automated tomography, the total accumulated electron dose following a large series of tilted images can now be restricted to 15 e$Å^{-2}$(Dirksen*et al.*, 19ⵓ3, 1995).

Examples of some specimen conditions that should be avoided during cryoelectron microscopy are shown in *Figures 7.17–7.20*. In the first example (*Figure 7.17*) contamination from large hexagonal ice crystals is

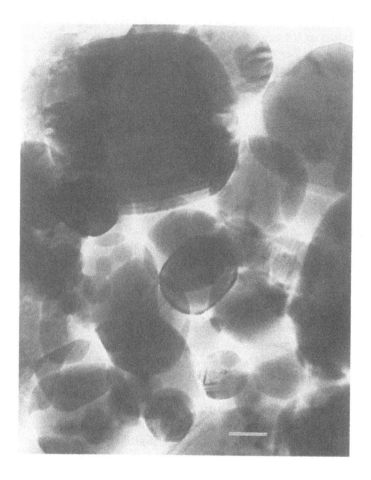

Figure 7.17: Crystalline ice contamination.
Part of a cryospecimen excessively contaminated by crystals of hexagonal ice. The scale bar indicates 100 nm. Micrograph courtesy of Marc Adrian.

shown; usually such regions can readily be avoided but the presence of smaller ice crystals is common (see *Figures 8.8* and *8.11–8.14*). In *Figure 7.18* a thick sheet of cubic ice can be seen to partly cross a hole also containing vitreous ice. Detail of the haemocyanin molecules present is clearly obliterated by the thicker cubic ice (but see Cyrklaff and Kühlbrandt, 1994). Excessive irradiation damage produces progressive bubbling of vitreous ice (*Figures 7.19* and *7.20*) and must always be avoided by observing strict low dose conditions.

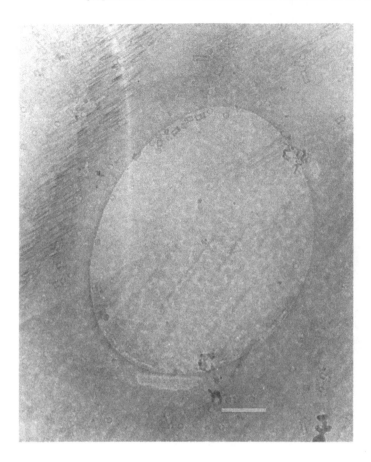

Figure 7.18: Cubic ice.
A hole in carbon support film, spanned by a layer of ice of varying thickness, indicating the presence of cubic rather than vitreous ice. The scale bar indicates 200 nm. Micrograph with the assistance of Shaoxia Chen and Helen Saibil.

7.5 Interpretation of cryoimages

The rather 'grainy' low contrast defocused images of unstained vitrified biological materials do contain structural information at the level of approximately 20–30 Å (Dubochet *et al.*, 1988). From crystalline materials, with correction for the phase-contrast transfer function (CTF), 3-D resolutions better than 10 Å have been achieved (Toyoshima and Unwin, 1988; Unwin, 1993, 1995). For single particle averaging, CTF correction is more difficult and has not yet been widely introduced, which restricts the resolution to about 20 Å for 2-D reconstructions and 30 Å for 3-D reconstructions. CTF correction has, however, been included for a number of viral 3-D reconstructions (Stewart *et al.*, 1993; Zhou *et al.*, 1996), from which

Figure 7.19: 'Bubbling'
Part of a region of thin vitreous ice spanning a hole in the specimen support film that has undergone an early stage of electron beam damage, showing the start of 'bubbling'. The scale bar indicates 100 nm. Micrograph courtesy of Marc Adrian.

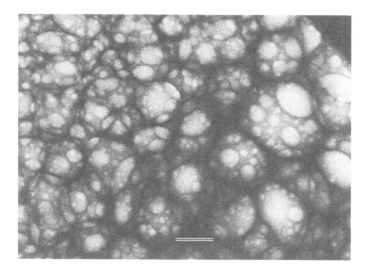

Figure 7.20: Extensive 'bubbling'.
Part of are region of vitreous ice spanning a hole, that has undergone electron beam damage, with extensive 'bubbling. The scale bar indicates 100 nm. Micrograph courtesy of Marc Adrian.

resolutions of 25 Å and better are now obtained (for further discussion see Chapter 10), and it is likely to rapidly become a standard part of single molecule image processing. It is thought that there will be a steady improvement of resolution in 3-D reconstructions from unstained cryoimages, so that for crystallographic analysis large well-ordered 2-D crystals will be increasingly required and for single particle analysis, data sets containing several thousand particles will be needed (see Henderson, 1995).

References

Battersby BJ, Sharp JCW, Webb RI, Barnes GT. (1994) Vitrification of aqueous suspensions from a controlled environment for electron microsocopy: an improved plunge-cooling device. *J. Microsc.* **176**, 110–120.

Bellare JR, Davis HT, Scriven LE, Talmon Y. (1986) An improved controlled-environment vitrification system (CEVS) for cryofixation of hydrated TEM samples. In *Proceedings of the XIth International Conference on Electron Microscopy* (eds T Imura, S Marus, T Suzuki) Vol II. Kyoto, pp. 367–368.

Bellare JR, Davis HT, Scriven LE, Talmon Y. (1988) Controlled environment vitrification system: an improved sample preparation technique. *J. Electr. Microsc. Technique* **10**, 87–111.

Berriman J, Unwin N. (1994) Analysis of transient structures by cryo-microscopy combined with rapid mixing of spray droplets. *Ultramicroscopy* **56**, 241–252.

Cyrklaff M, Kühlbrandt W. (1994) High-resolution electron microscopy of biological specimens in cubic ice. *Ultramicroscopy* **55**, 141–153.

Cyrklaff M, Roos N, Gross H, Dubochet J. (1994) Particle-surface interaction in thin vitrified films for cryo-electron microscopy. *J. Microsc.* **175**, 135–142.

Dirksen K, Typke D, Hegerl R, Baumeister W. (1993) Towards automatic electron tomography. II. Implementation of autofocus and low-dose procedures. *Ultramicroscopy* **49**, 109–120.

Dirksen K, Typke D, Hegerl R, Walz J, Sackman E, Baumeister W. (1995) Three-dimensional structure of lipid vesicles embedded in vitreous ice and investigated by automated electron tomography. *Biophys J.* **68**, 1416–1422.

Dubochet J, Adrian M, Chang J-J, Homo J-C, Lepault J, McDowall AW, Schulz P. (1988) Cryo-electron microscopy of vitrified specimens. *Quart. Rev. Biophys.* **21**, 129–228

Frederick PM, Stuart MCA, Bomans PHH, Busing WM, Burger KLJ, Verkleij AJ. (1991) Perspective and limitations of cryo-electron microscopy. *J. Microsc.* **161**, 253–262.

Henderson R. (1995) The potential and limitations of neutrons, electrons and X-rays for atomic resolution microscopy of unstained biological molecules. *Quart. Rev. Biophys.* **28**, 171–193.

Siegel DP, Green WJ, Talmon Y. (1994) The mechanism of lamellar-to-inverted hexagonal phase transition: a study using temperature-jump cryo-electron microscopy. *Biophys. J.* **66**, 402–414

Stewart PL, Fuller SD, Burnett RM. (1993) Difference imaging of adenovirus: bridging the resolution gap between X-ray crystallography and electron microscopy. *EMBO J.* **12**, 2589–2599.

Subramanian S, Gerstein M, Osterhelt D, Henderson R. (1993) Electron diffraction analysis of structural changes in the photocycle of bacteriorhodopsin. *EMBO J.* **12**, 1–8.

Toyoshima C, Unwin N. (1988) Contrast transfer for frozen-hydrated specimens: determination from pairs of defocused images. *Ultramicroscopy* **25**, 279–292.

Trinick J, Cooper J. (1990) Concentration of solutes during preparation of aqueous suspensions for cryo-electron microscopy. *J. Microsc.* **159**, 215–222.

Unwin PNT. (1993) Nicotinic acetylcholine receptor at 9 Å resolution. *J. Mol. Biol.* **229**, 1101–1124.

Unwin PNT. (1995) Acetylcholine receptor channel image in the open state. *Nature* **373**, 37–43.

Walker M, White H, Belknap B, Trinick J. (1994) Electron cryomicroscopy of acto-myosin-S1 during steady-state ATP hydrolysis. *Biophys. J.* **66**, 1563–1572.

Zhou ZH, Hardt S, Wang B, Sherman MB, Jakana J, Chiu W. (1996) CTF determination of images of ice-embedded single particles using a graphics interface. *J. Struct. Biol.* **116**, 216–222.

8 Unstained Vitrified Specimens: Selected Applications

The steadily increasing number of biological applications of cryoelectron microscopy of unstained vitrified specimens has made the selection of topics and representative examples difficult. As with negative staining (Chapter 5) the space available for applications together with their structural and biological interpretation is restricted and inevitably the selection will be incomplete. However, the interested reader should easily find relevant publications in the literature and initially the surveys by Dubochet *et al.* (1986) and Stewart (1991) should be useful. Thin film cryoelectron microscopy has been applied almost entirely to isolated systems, rather than intact cells and organelles. However, one example which parallels negative staining studies, is the application of high voltage cryoelectron microscopy to the study of platelet structure (O'Toole *et al.*, 1993).

The contents of this chapter are very much related to the material included in Chapter 9 (future prospects), Chapter 10 (image processing) and Chapter 5 (negative staining applications). It is encouraging that most TEM images of unstained biological material embedded in a thin film of vitreous ice closely corresponded to their equivalent negatively stained images. With the cryoimages it has become essential that image processing with 2-D or 3-D image reconstruction should be performed, since the low image contrast does not always present the eye with sufficient information to fully appreciate the structures under investigation. This may even be the case with appropriately increased contrast and signal to noise induced by under focus, since the noise level of the image usually remains relatively high. Thus, several examples including image processing will be given here, and later (Chapters 9 and 10).

8.1 Lipids and membranes

In recent years there have been numerous cryoelectron microscopical applications to lipids, membranes and lipoproteins. Unstained images of cholesterol microcrystals obtained by cryoelectron microscopy (*Figure 8.1*) correlate well with those obtained by negative staining and metal shadowing (see Harris, 1988, and cf. *Figure 5.4*). It is significant that the multi-

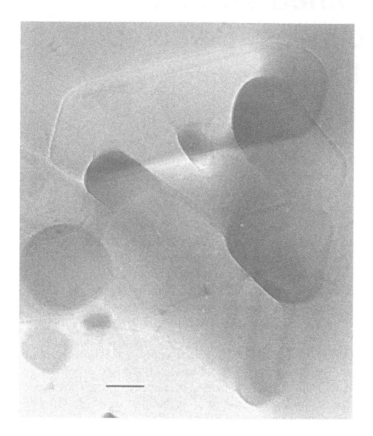

Figure 8.1: Vitrified microcrystals of cholesterol.
A cryoelectron micrograph of unstained vitrified microcrystals of cholesterol (see *Figure 5.4* for images of equivalent material in negative stain). The scale bar indicates 100 nm. Micrograph courtesy of Marc Adrian.

bilayer structure of such material can be defined at the edge of some of the crystals, despite the low image contrast. Crystalline cholesterol presents a useful surface upon which dynamic/time-dependent interactions of cholesterol-binding toxins could be monitored following rapid freezing. This in turn may assist the detailed understanding of toxin-binding and subsequent lesion formation (see also Chapter 5, *Figures 5.5* and *5.6)*.

A considerable number of cryoelectron microscopic studies have been devoted to the investigation of liposomes. This interest stems from the more fundamental biophysical level where information can be gained on lipid phase changes, bilayer fusion and protein–lipid interactions (Klösgen *et al.*, 1993; Siegel *et al.*, 1994; Tahara and Fujiyoshi, 1994), through to more applied aspects of interest to the pharmaceutical industry where liposomes are used as carriers (microcapsules) for a range of different drugs. An example showing predominantly single bilayer liposomes of graded size, is given in *Figure 8.2*. It is significant that the larger liposomes have located themselves in the thicker fluid film, towards the edge of the hole in the carbon support, prior to rapid freezing. This phenomenon is often encountered with unstained material supported across these holes, together with the clustering of particles at the hole edge, it must be accepted as a natural occurrence which is sometimes beneficial.

The nuclear envelope from the amphibian oocyte has been studied thoroughly by cryoelectron microscopy in recent years and this has led to a considerable advance in the understanding of the structure and function of the nuclear pore complex. *Figure 8.3* shows part of a spread nuclear

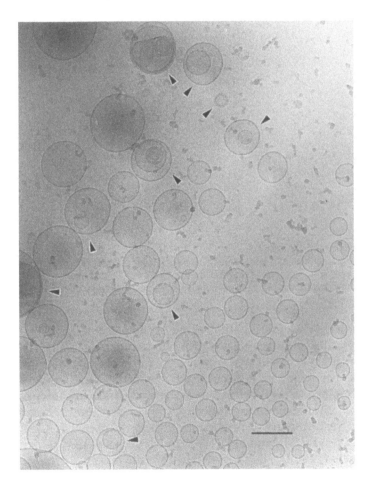

Figure 8.2: Unstained vitrified liposomes.

A cryoelectron micrograph of unstained vitrified liposomes. Most of the liposomes present possess a single bilayer, but a few multi-bilayer structures are present (arrows). Note the gradation of liposome size, in parallel with the progressively thicker vitreous ice film (top left) moving towards the edge of the hole in the carbon support film. The scale bar indicates 200 nm. Micrograph courtesy of Jacques Dubochet.

Figure 8.3: Nuclear envelope in vitreous ice.
A cryoelectron micrograph of part of a spread unstained nuclear envelope from a *Necturus* oocyte in vitreous ice. Note the octagonal symmetry of the NPCs and the presence of the central transporter in many instances. (Cf. *Figure 5.14a.*) The scale bar indicates 100 nm. Micrograph courtesy of Christopher Akey.

envelope from a *Necturus* oocyte. In this region of the nuclear envelope, the nuclear pore complexes form a partly ordered tetragonal array; it is not known whether this is of structural significance in relation to the symmetry of the nuclear pore complex. The central transporter of a number of the nuclear pore complexes is well defined as is the octagonal symmetry of the annular material. Single particle analysis of the nuclear pore complex has been performed, resulting in 2-D projection averages (*Figure 8.4*; Akey, 1990; cf. *Figures 5.12* and *5.13*), and also 3-D reconstructions (Akey, 1995; Akey and Radermacher, 1993). Such reconstructions do, however, need to be carefully assessed alongside information from other approaches such as metal shadowing TEM, high resolution SEM and dark field STEM analysis, as well as from negative staining TEM.

The cryoelectron microscopy of 2-D crystals of membrane proteins, and also soluble proteins adsorbed as 2-D crystals on to lipid monolayers, continues to be a major area of expansion (Asturias and Kornberg, 1995; Böttcher *et al.*, 1995; Brisson *et al.*, 1994; Guo *et al.*, 1995; Henderson *et al.*, 1990; Jap and Li, 1995; Kühlbrandt *et al.*, 1994; Stokes and Green 1990; Unger and Schertler, 1995). These studies, often in combination with electron diffraction analysis, have produced the highest resolution electron microscopical 2-D projections and 3-D macromolecular reconstructions thus far obtained. However, it should be emphasized that many of these studies with glucose, trehalose and tannic acid as embedding media for 2-D crystals utilized specimen cooling within the TEM to −170°C or below, rather than the actual production and cryotransfer of frozen-hydrated vitreous specimens.

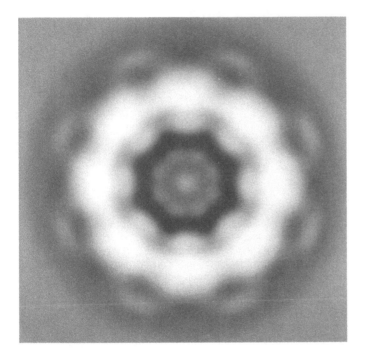

Figure 8.4: Nuclear pore complex in vitreous ice.
   An average projection map of a single unstained nuclear pore complex in vitreous ice, with the central transporter composed of eight peripheral density peaks which surround a central peak and a dark ring which may represent the edge of the expanded transport pore (cf *Figure 5.14b*, for a negatively stained NPC rotational average). Reconstruction courtesy of Christopher Akey. Reproduced from Akey (1990) Visualization of transport-related configurations of the nuclear pore transporter. *Biophys. J.* 58, 341–355, with permission fom the Biophysical Society.

8.2   Tubular and filamentous structures

The structural analysis of muscle actin and myosin filaments has progressed significantly due to the impact of cryoelectron microscopy (Jontes *et al.*, 1995; Menetret *et al.*, 1990; Schröder *et al.* 1993, Sosinsky *et al.*,

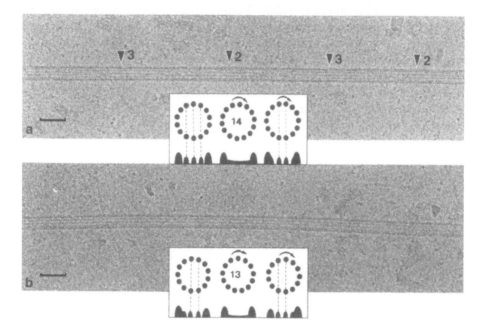

Figure 8.5: Microtubules in vitreous ice.

Images of (a) 14- and (b) 13-protofilament unstained microtubules in vitreous ice. The inset drawings show how the major features in the images are produced by projecting the protofilament structure along the electron beam direction. The strong edge contrast is due to superimposition of several protofilaments in projection and the finer inner fringes occur when a protofilament on the top surface projects in register with a protofilament below. For an even number of protofilaments the image will always be symmetric whilst for odd numbers the fringe contrast will be offset towards one of the edges. The 14-protofilament microtubule shows repetitions of the characteristic three fringe/blurred/two fringe motif (arrowed). A characteristic of 13-protofilament microtubules is the constant contrast over considerable lengths. The scale bars indicate 50 nm. Micrographs courtesy of Dick Wade. Reproduced from Wade and Chrétien (1993) Cryoelectron microscopy of microtubules. *J. Struct. Biol.* 110, 1–27, with permission from Academic Press.

1992; Walker *et al.*, 1995; Whittaker *et al.*, 1995). Often these studies have included helical image processing and 3-D image reconstruction, aimed at defining the protein–protein interactions of importance for filament formation and muscle function. It is likely that the other filamentous proteins such as intermediate filaments, the Alzheimer (tau and paired helical filaments) proteins (Mandelkow *et al.*, 1995), bacterial pili and flagellae (DeRosier, 1995; Morgan *et al.*, 1995) will all continue to receive attention from cryoelectron microscopists. Similarly, the understanding of microtubule dynamics and protofilament organization has benefited significantly from cryoelectron microscopy (Arnal and Wade, 1995; Mandelkow *et al.*, 1991; Wade and Chrétien, 1993). The example given in *Figure 8.5* shows how cryoelectron microscopy has provided an assessment of the number of tubulin protofilaments in microtubules of slightly different diameter.

The successful cryoelectron microscopy of tannic acid-stabilized 2-D crystals of tubulin (Nogales *et al.*, 1995a, b) indicates that this approach may also help in the understanding of microtuble organization.

The cryoelectron microscopy of nucleic acid and nucleoprotein complexes has been intensively pursued by Jacques Dubochet and his colleagues for a number of years (Dubochet *et al.*, 1992, 1994; Furrer *et al.*, 1995; Gogol *et al.*, 1991). An example of the pUC9 2673 base pair DNA from *E. coli* is shown in *Figure 8.6*. Regions of the flexible DNA molecule that are more tightly supercoiled can be defined.

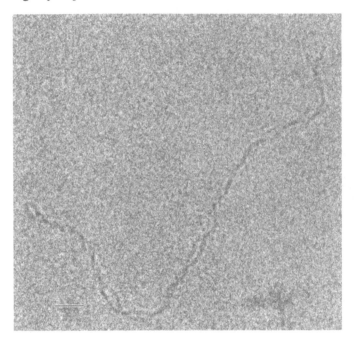

Figure 8.6: The pUC9 2673 base pair DNA from *E. coli* in vitreous ice.
A cryoelectron micrograph of the pUC9 2673 base pair DNA from *E. coli*, unstained, in vitreous ice. Regions where the closed DNA loop is more tightly and loosely supercoiled can be defined. The scale bar indicates 50 nm. Micrograph courtesy of Jacques Dubochet (cf. Adrian *et al.*, 1990; ten Heggeler-Bordier *et al.*, 1992).

8.3 Macromolecules and macromolecular assemblies

Recently, much effort has been devoted to the cryoelectron microscopy of many soluble macromolecules and macromolecular assemblies (Chen *et al.*, 1994; Frank *et al.*, 1995; Spin and Atkinson, 1995), often emphasizing the agreement between the cryodata and that from negative staining (Stoops *et al.*, 1991). An interesting example, containing a mixture of horse spleen ferritin and chemically produced apoferritin is shown in *Figure 8.7*

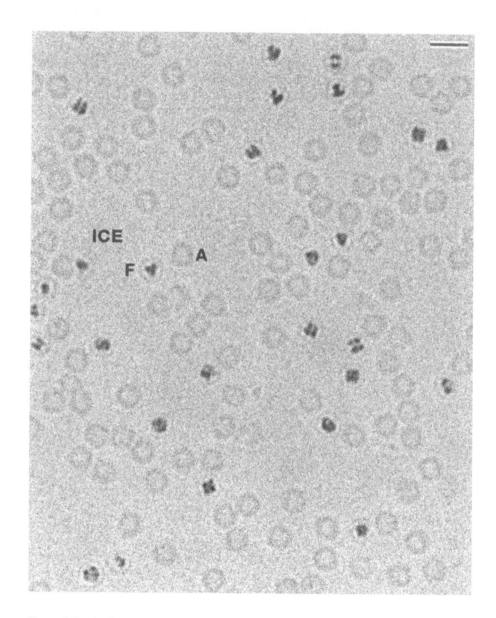

Figure 8.7: A mixture of ferritin and apoferritin in vitreous ice.

A cryoelectron micrograph of a mixture of unstained horse spleen ferritin and chemical apoferritin in vitreous ice. The iron hydroxide core of the ferritin molecules (F) has a high image contrast and shows a notably diverse morphology; apoferritin molecules (A) have an empty central cavity filled with vitreous ice (ICE). Phase-contrast granularity in the vitrified ice is nearly absent despite the use of a large defocus needed to image the unstained protein. The thickness and symmetry of the protein shell in ferritin molecules is directly related to the size, shape and position of the central core; note the bright Fresnel fringe immediately surrounding each core. The scale bar indicates 20 nm. Micrograph courtesy of Bill Massover and Marc Adrian. Reprinted from Massover (1993) Ultrastructure of ferritin and apoferritin: a review. *Micron* 24, 389–437, with permisson from Elsevier Science Ltd.

(cf. *Figure 5.33*). At the level of instrumental defocus selected the phase-contrast granularity of the image is not excessive, but the iron hydroxide cores of the molecules produce a bright Fresnel fringe that interferes with the *apparent* thickness of the protein shell when compared to that of the iron-deficient apoferritin molecules (Massover, 1986, 1993; Massover and Adrian, 1986). Our own studies on the structure of molluscan haemocyanin have recently benefited from cryoelectron microscopy (Dube *et al.*, 1995) and a low resolution (approx. 45 Å) 3-D reconstruction has been produced, as shown in *Figures 8.8* and *8.9* (see also Chapters 5 and 9; Lambert *et al.*, 1994, 1995a,b). Parallel studies have also been performed using cryoelectron microscopy to obtain low resolution 3-D reconstructions from the high molecular weight annelid haemoglobin/erythrocruorin (de Haas *et al.*, 1996; Schatz *et al.*, 1995).

Usually in cryoelectron microscopy, the frozen hydrated specimens are supported across the holes of carbon film, although if the carbon itself is reasonably thin, the deterioration of the signal to noise may not be great (see *Figure 8.8a*). Thus, single molecules, 2-D arrays and crystals adsorbed on to a thin continuous carbon support film during the NS–CF procedure (Chapter 3.5) can be transferred to distilled water and rapidly frozen (*Figure 8.10*); possibly, this technology could be developed for the future cryoelectron microscopy of 2-D crystals of soluble proteins, viruses and 'wet' cleaved cells. Use of an extremely thin carbon support film, subsequently supported across a thicker holey carbon film (as was proposed within the original NS–CF procedure; Horne and Pasquali-Ronchetti, 1974) is also likely to be useful. Additionally, the possibility of creating 2-D crystals directly upon a *non-adsorptive* holey carbon support film instead of on a mica surface might extend this approach more appropriately for rapid freezing and cryoelectron microscopy (see also Chapter 9).

Our on-going studies on the reassociation of keyhole limpet haemocyanin from purified subunits have also been supplemented by cryoelectron microscopic studies (in collaboration with Marc Adrian), and have been performed in parallel with room temperature and cryonegative staining. A bundle of haemocyanin tubules is shown in *Figure 8.11*, following reassociation/polymerization of the subunit from KLH1 in the presence of a high concentration of calcium and magnesium ions. When compared to the same sample material imaged by negative staining (*Figure 5.24*) a very good agreement between the two specimen preparations is found. The oblique structural feature of the individual flexible tubular polymers, indicative of incompletely annealed helical tubules (i.e. an open, twisted ribbon structure) is present in both these figures. Examples of reassociated KLH2 (containing a biochemically and immunologically different polypeptide to KLH1) which forms a tubular polymer and stacked decamer structures (multidecamers) are shown in *Figures 8.12* and *8.13*. In the former example, the vitreous ice is somewhat thinner, revealing the banded repeating structure of the multidecamers alongside the tubule structures (which are significantly different to those of KLH1; cf. *Figure 8.11*). The

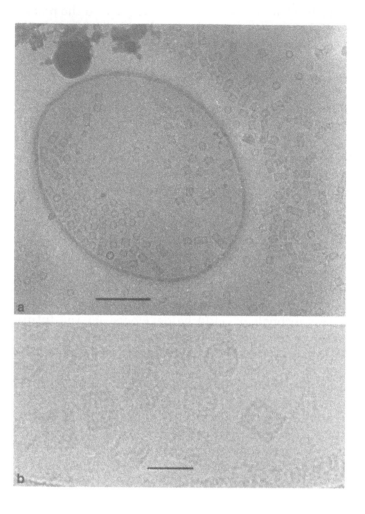

Figure 8.8: KLH in vitreous ice.

Cryoelectron micrographs of unstained total (unfractionated) KLH in vitreous ice. (a) With the assistance of Shaoxia Chen and Helen Saibil; (b) courtesy of Marin van Heel and Prakash Dube. The scale bars indicate 200 nm (a) and 50 nm (b). Part (b) reproduced from Dube *et al.* (1995) Three-dimensional structure of keyhole limpet hemocyanin by cryoelectron microscopy and angular reconstitution. *J. Struct. Biol.* 115, 226–232, with permission from Academic Press.

wide range of particle size and diverse structural information present in *Figure 8.12* can be readily appreciated. The slight loss of apparent detail exhibited by *Figure 8.13* (because of the thicker film of vitreous ice) is more than compensated for by the greater range of particle orientation seen (due to a greater freedom of mobility in the slightly thicker aqueous film immediately prior to rapid freezing). This indicates that a very thin layer of fluid containing particulate material when spread across the holes of the holey carbon support film restricts macromolecular orientation, as considered previously by Dubochet *et al.* (1985, 1986). Undoubtedly, this is an extremely important point for the retrieval of 3-D information for image processing, where a full range of particle orientation is desirable.

Immunoelectron microscopical studies (cf. *Figures 5.36–5.40*) on immune complexes of varying size (groups of molecules specifically linked by monoclonal antibodies) have yet to be pursued intensively by cryoelectron microscopy (but see Boisset *et al.*, 1995). Since such complexes are 3-D structures in solution, they will tend to flatten to some extent in negatively stained specimens (but see Chapter 9; negative staining on holey carbon films). Cryoelectron microscopy, when a relatively thick aqueous film has been vitrified, provides the possibility of retaining the three-dimensionality of small immune complexes; I think that in the future there

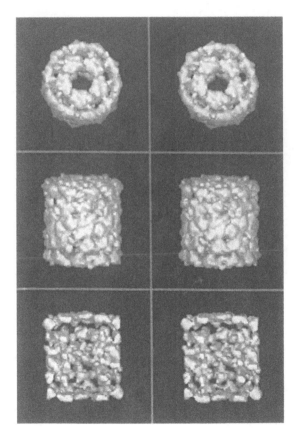

Figure 8.9: 3-D image reconstruction of the KLH didecamer.
Stereo pairs of the c. 45 Å 3-D image reconstruction of the KLH didecamer present in *Figure 8.8b*. Note the five 'collar' elements showing clearly in the end view (top pair), the surface helical feature of the didecamer in side view (middle pair) and the considerable projection of protein into the central cavity (lower pair). Photograph courtesy of Marin van Heel and Prakash Dube. Modified from Dube *et al.* (1995) Three-dimensional structure of keyhole limpet hemocyanin by cryoelectron microscopy and angular reconstitution. *J. Struct. Biol.* 115, 226–232.

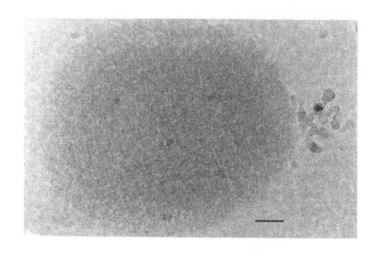

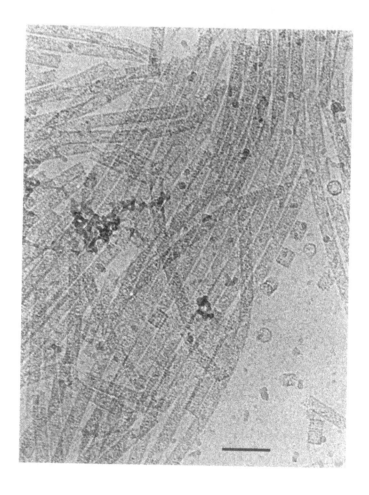

◀ **Figure 8.10:** Vitrified *Haliotis* haemocyanin didecamers.
A cryoelectron micrograph showing a 'pool' of close-packed unstained vitrified *Haliotis* haemocyanin didecamers, produced by the NS–CF procedure with transfer to *distilled water* before vitrification (i.e. all ammonium molybdate washed away). Note that in this instance the protein molecules are supported on a thin *continuous* carbon film. The scale bar indicates 100 nm. Micrograph with the assistance of Helen Saibil and Shaoxia Chen.

will be a steady expansion of immunoelectron microscopical studies at the molecular level (Boisset *et al.*, 1995) using antibodies and gold cluster-labelled antibodies (as well as gold cluster labelling of functional groups on macromolecules; Boisset *et al.*, 1992; Wagenknecht *et al.*, 1994). Both cryoelectron microscopy and negative staining (with low density stains) are likely to continue to play a part in such investigations; with antibody alone (IgG, Fab, Fab') and with the 1.4 nm Nanogold- and 0.8 nm undecagold–antibody or streptavidin conjugates (available from Nano-probes Inc., Stony Brook, NY). Thus, considerable future improvement in the high resolution definition of epitope location(s) on macromolecules and viruses can be anticipated.

8.4 Viral structure

Perhaps due to the inherent complexity of viral structure and the relatively slow progress of X-ray crystallography of viruses (despite the impressive X-ray studies of Michael Rossmann and his colleagues; Agbanje *et al.*, 1994; McKenna*et al.*, 1994; Rossmann, 1994; Rossmann*et al.*, 1994) viral structure has continued to be a principal and expanding domain for cryoelectron microscopical investigations. However, the X-ray crystallography of individual viral proteins has advanced rapidly and has provided atomic structures (from the proteins alone and as complexes with antibody or anti-viral compounds), some of which have implications for 'peptide fitting' of individual proteins within lower resolution cryoelectron microscopic 3-D reconstructions of viruses (Stewart *et al.*, 1993).

Of the large quantity of available cryoelectron microscopical data on animal, plant and bacterial viruses, relatively few examples can be presented here. Some of the earliest cryoelectron microscopy was performed on viruses (Adrian *et al.*, 1984; Lepault, 1985; Vogel *et al.*, 1986) indicating the potential of this approach for the study of viral structure. The increasing availability of software for the image processing of such data in

◀ **Figure 8.11:** Vitrified bundle of the KLH1 tubular polymer.
Cryoelectron micrograph of an unstained vitrified disordered bundle of the KLH1 tubular polymer, produced by reassociation of KLH1 subunits in the presence of 100 mM calcium and magnesium chloride. Note the presence of a few didecamers and the oblique helical feature of the tubules, also shown by negatively stained images (see *Figure 5.24*). The scale bar indicates 100 nm. Micrograph courtesy of Marc Adrian.

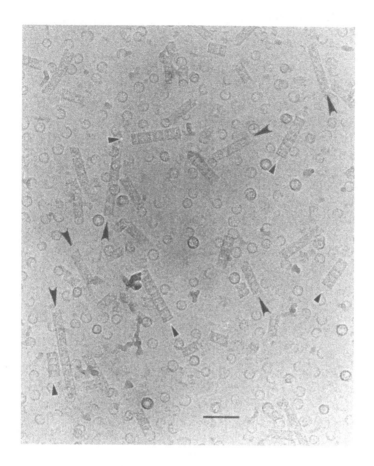

Figure 8.12: Reassociated KLH2 in thin vitreous ice.

A cryoelectron micrograph of KLH2, reassociated in the presence of 100 mM calcium and magnesium chloride (cf. Harris *et al.*, 1995) showing a region of thin vitreous ice (cf. *Figure 8.13)*. Note the presence of small, partly formed, KLH2 ring-like oligomers, stacked-disk multidecamers (small arrows) and smaller diameter tubular polymer (large arrows). The scale bar indicates 100 nm. Micrograph courtesy of Marc Adrian.

the form of single viruses at varying orientation within the image, leading to 3-D reconstructions (see Chapter 10), has led to the widespread use of cryoelectron microscopy in the field of structural virology (Adrian *et al.*, 1992; Baker *et al.*, 1991; Fuller, *et al.*, 1995; Hewat, *et al.*, 1992, 1994; Metcalf *et al.*, 1991; Newcomb *et al.*, 1993; Paredes *et al.*, 1993; Shaw *et al.*, 1993; Smith *et al.*, 1995; Wang *et al.*, 1992; Zhou *et al.*, 1994).

The *classical* EM image of human and simian immunodeficiency viruses (HIV and SIV) and other lentiviruses, with a membrane-enveloped conical core, derived from thin sectioning studies is well known by all (Grief *et al.*, 1994; Harris, 1993). The rather irregular shape of the lentiviruses has been supported by both negative staining and cryoelectron

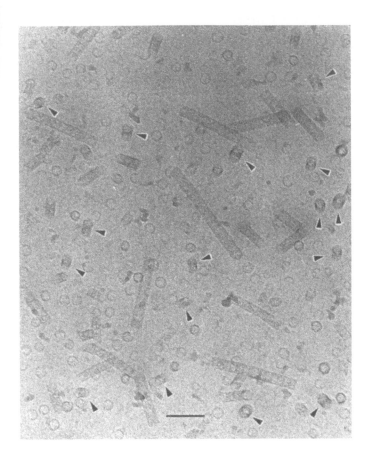

Figure 8.13: Reassociated KLH2 within a thick film of vitreous ice.

A cryoelectron micrograph of reassociated KLH2, as in *Figure 8.12*, but showing a region possessing a considerably thicker film of vitreous ice. Note the large number of indistinct tilted images (arrows) of the short multidecamers and tubules, indicating that the freedom of mobility of the oligomeric and polymeric structures is increased within the thicker aqueous film immediately prior to vitrification. The scale bar indicates 100 nm. Micrograph courtesy of Marc Adrian.

microscopy (Nermut *et al.*, 1993), see *Figure 8.14*. Because of the considerable image variability, computer processing has yet to be performed on HIV or SIV. It is, however, appropriate to mention here that the complex pentagonal organization of the HIV–gag shell protein, revealed by negative staining with sodium silicotungstate (Nermut *et al.*, 1994), has been confirmed by high resolution X-ray diffraction studies on the SIV matrix antigen/virion shell protein (Rao *et al.*, 1995).

Figure 8.15 shows one of the early cryoelectron microscopical images of adenovirus type 2 (courtesy of Jacques Dubochet), presented in reverse contrast for direct comparison with the negatively stained images of adenovirus (*Figures 5.24–5.26*). The viral capsid hexons and the penton

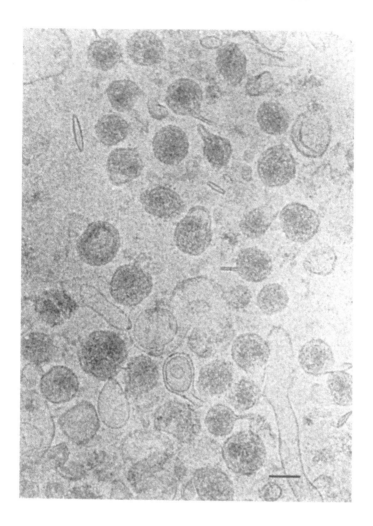

Figure 8.14: Vitrified simian immunodeficiency virus (SIV).

A cryoelectron micrograph of unstained vitrified SIV, prepared on a thin carbon film. Note the considerable variability of the viral core shape and the viral membrane of this lentivirus, in agreement with data from thin sectioning. Modified from Nermut *et al.* (1993). The scale bar indicates 100 nm. Micrograph courtesy of Milan Nermut and Frank Booy.

spikes show very clearly . Such images have led to the impressive computer processing and image reconstruction of adenovirus performed by Roger Burnett, Steve Fuller and their colleagues (Stewart and Burnett, 1995; Stewart *et al.*, 1991, 1993; see also Chapter 10). Interestingly, an earlier dark-field STEM study of the group of nine hexons (Furcinitti *et al.*, 1989) made a useful contribution towards the localization of a small viral protein (polypeptide IX) in the spaces between the capsid hexons. Analysis of the adenovirus penton fibre has also been performed by cryoelectron microscopy of small 3-D crystals (Devaux *et al.*, 1990).

As with negative staining, the production of unstained specimens in vitreous ice can generate some adenovirus disruption (see *Figure 8.16*; cf. *Figure 5.24*). Again, the group of nine (GON) hexons is found to break away from the viral capsid, but in this instance only GON hexons possessing the left-hand orientation are present. This result is highly suggestive of viral disruption at the air–fluid interface, with removal of (all) the group of nine hexons possessing the right-hand orientation at the time of specimen blotting, immediately prior to freezing. (Dubochet *et al.*, 1985, 1986).

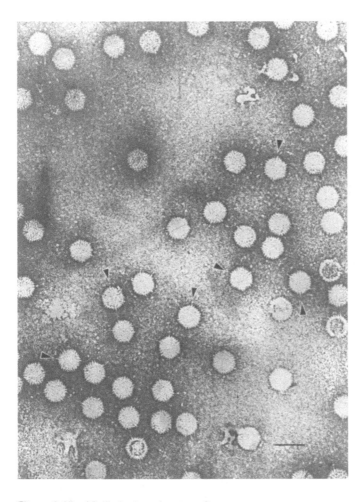

Figure 8.15: Vitrified adenovirus type 2.
A cryoelectron micrograph of unstained vitrified adenovirus type 2 (contrast reversed). Note the uniform distribution of the viruses and their well-defined shape. The superimposition of the upper and lower side of the viruses makes the surface structure difficult to interpret. Spikes are visible in favourable case (arrows) (see images of negatively stained adenovirus, *Figure 5.26 – 5.29*). The scale bar indicates 100 nm. Micrograph courtesy of Jacques Dubochet. Reprinted with permission from Adrian *et al.* (1984) Cryo-electron microscopy of viruses. *Nature* 308, 32–36. Copyright 1984 Macmillan Magazines Ltd.

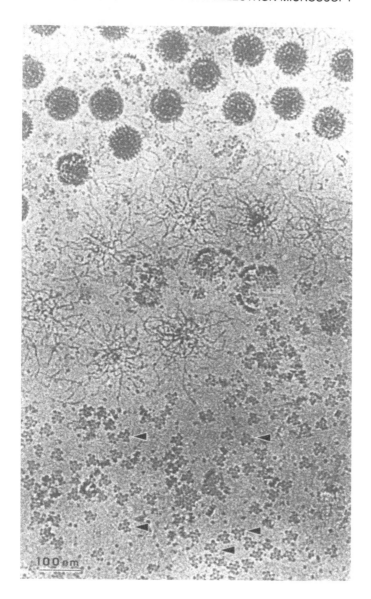

Figure 8.16: Vitrified adenovirus type 2 exhibiting partial disruption.

A cryoelectron micrograph of unstained vitrified adenovirus type 2. In this instance there is partial disruption of some of the viruses. Note the release of nucleic acid (centre) and groups of nine (GON) hexons (arrows). The GON are all in the left-hand orientation, because of the selective removal of those in the right-hand orientation during blotting (see negatively stained GONs, *Figure 5.27*). Micrograph courtesy of Jacques Dubochet. Reprinted from Dubochet *et al.* (1985) Cryo-electron microscopy of vitrified biological specimens. *Trends Biochem. Sci.* 10, 143–146, with permission from Elsevier Science Ltd.

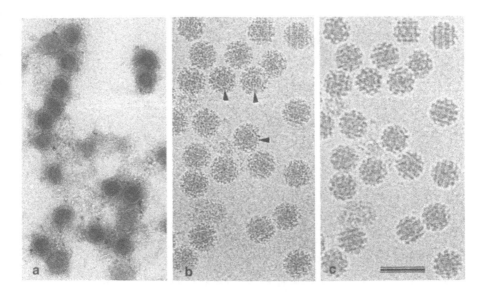

Figure 8.17: Semliki Forest virus (SFV).
Semliki Forest virus, (a) negatively stained with 1% uranyl acetate, (b and c) unstained cryoelectron micrographs at 3 μm and 8 μm defocus, respectively. Note the indication of the lipid bilayer (arrows, b) and in (a) apparently due to shrinkage of the virus in the acidic uranyl acetate. The scale bar indicates 100 nm. Micrographs courtesy of Steve Fuller.

Semliki Forest virus (SFV), an enveloped RNA virus, has also received considerable attention from cryoelectron microscopists (Adrian *et al.*, 1984; Vogel *et al.*, 1986). An interesting feature of this virus is that the viral spikes show very clearly, yet it has proved to be difficult to resolve the lipid bilayer. The lipid bilayer, surrounding the nucleocapsid, can be seen when the level of instrumental defocus is not too great, as shown in *Figure 8.17* (see also Vogel *et al.*, 1986). This lipid bilayer can, however, be defined by negative staining with uranyl acetate, when the virus undergoes stain-induced shrinkage (*Figure 8.17a*). If close packed within a very thin aqueous film, viral particles, such as SFV may create a close packed monolayer, as shown in *Figure 8.18*. Although at first glance there is an impression of short-range order, in fact such monolayers do contain a close-packed array with viruses in apparently random orientations; they are not 2-D crystals. Nevertheless, the production of these viral monolayers indicates that there is a physical parallel between such phenomena, produced at the fluid–air interface (Pum *et al.*, 1993) and the production of 2-D arrays and crystals on mica by the NS–CF procedure, apparently also at the fluid–air interface (see also Chapters 3 and 5).

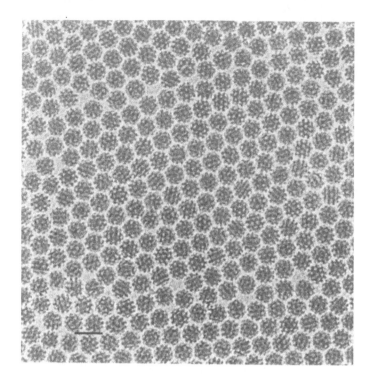

Figure 8.18: Two-dimensional array of vitrified Semliki Forest virus (SFV).
A cryoelectron micrograph of unstained vitrified SFV. Note that within the close-packed 2-D array the viruses are orientated in many different positions. The scale bar indicates 100 nm. Micrograph courtesy of Jacques Dubochet.

References

Adrian M, Dubochet J, Lepault J, McDowall AW. (1984) Cryo-electron microscopy of viruses. *Nature* **308**, 32–36.

Adrian M, ten Heggeler-Bordier B, Wahli W, Stasiak AZ, Stasiak A, Dubochet J. (1990) Direct visualization of supercoiled DNA molecules in solution. *EMBO J.* **9**, 4551–4554.

Adrian M, Timmins PA, Witz J. (1992) *In vitro* decapsidation of turnip yellow mosaic virus investigated by cryo-electron microscopy: a model for the decapsidation of a small isometric virus. *J. Gen. Virol.* **73**, 2079–2083.

Agbandje M, Kajigaya S, McKenna R, Young NS, Rossmann MG. (1994) The structure of human parvovirus B19 at 8 Å resolution. *Virology* **203**, 106–115.

Akey CW. (1990) Visualization of transport-related configurations of the nuclear pore transporter. *Biophys. J.* **58**, 341–355.

Akey CW. (1995) Structural plasticity of the nuclear pore complex. *J. Mol. Biol.* **248**, 273–293.

Akey CW, Radermacher M. (1993) Architecture of the *Xenopus* nuclear pore complex revealed by three-dimensional cryo-electron microscopy. *J. Cell Biol.* **122**, 1–19.

Arnal I, Wade RH. (1995) How does taxol stabilize microtubules? *Curr. Biol.* **5**, 900–908.

Asturias FJ, Kornberg RD. (1995) A novel method for transfer of two-dimensional crystals from the air/water interface to specimen grids. *J. Struct. Biol.* **114**, 60–66.

Baker TS, Newcombe WW, Olson NH, Cowser LM, Olson C, Brown JC. (1991) Structures of bovine and human papillomaviruses. *Biophys. J.* **60**, 1445–1456.

Boisset N, Grassucci R, Penczek P, Delain E, Pochon F, Frank J, Lamy JN. (1992) Three-dimensional reconstruction of a human α_2-macroglobulin with monomaleimido Nanogold (Au_{1-4} nm) embedded in ice. *J. Struct. Biol.* **109**, 39–45.

Boisset N, Penczek P, Taveau J-C, Lamy J, Frank J, Lamy J. (1995) Three-dimensional reconstruction of *Androctonus australis* hemocyanin labeled with a monoclonal Fab fragment. *J. Struct. Biol.* **115**, 16–29.

Böttcher B, Gräber P, Boekema EJ, Lücken U. (1995) Electron cryomicroscopy of two-dimensional crystals of the H^+–ATPase from chloroplasts. *FEBS Lett.* **373**, 262–264.

Brisson A, Olofsson A, Ringler P, Schmutz M, Stoylova S. (1994) Two-dimensional crystallization of proteins on planar lipid films and structure determination by electron crystallography. *Biol. Cell* **80**, 221–228.

Chen S, Roseman AM, Hunter AS, Wood SP, Burston SG, Ranson NA, Clarke AR, Saibil HR. (1994) Location of a folding protein and shape changes in GroEL–GroES complexes imaged by cryo-electron microscopy. *Nature* **371**, 261–264.

DeRosier DJ. (1995) Spinning tails. *Curr. Opin. Struct. Biol.* **5**, 187–193.

Devaux C, Adrian M, Berthet-Colominas C, Cusack S, Jacrot B. (1990) Structure of Adenovirus fibre: I. Analysis of crystals of fibre from adenovirus serotypes 2 and 5 by electron microscopy and X-ray crystallography. *J. Mol. Biol.* **215**, 567–588.

Dube P, Orlova EV, Zemlin F, van Heel M, Harris JR, Markl J. (1995) Three-dimensional structure of keyhole limpet hemocyanin by cryoelectron microscopy and angular reconstitution. *J. Struct. Biol.* **115**, 226–232.

Dubochet J, Adrian M, Lepault J, McDowall AW. (1985) Cryo-electron microscopy of vitrified biological specimens. *Trends Biochem. Sci.* **10**, 143–146.

Dubochet J, Adrian M, Chang J-J, Homo J-C, Lepault J, McDowall AW, Schultz P. (1986) Cryoelectron microscopy of vitrified specimens. *Quart. Rev. Biophys.* **21**, 129–228.

Dubochet J, Adrian M, Dustin I, Furrer P, Stasiak A. (1992) Cryoelectron microscopy of DNA molecules in solution. In *Methods in Enzymology*, Vol. 211, DNA structures, Part A, Synthesis and physical analysis of DNA (eds DMJ Lilley, JE Dahlberg). Academic Press, San Diego, pp. 507–518.

Dubochet J, Bednar J, Furrer P, Stasiak A. (1994) Spatial visualization of DNA in solution. *J. Struct. Biol.* **107**, 15–21.

Frank J, Zhu J, Penczek P, Liu Y, Srivastava S, Verschoor A, Radermacher M, Grassucci R, Lata RK, Agrawal RK. (1995) A model of protein synthesis based on cryo-electron microscopy of the *E. coli* ribosome. *Nature* **376**, 441–444.

Fuller SD, Berriman JA, Butcher SJ, Gowen BE. (1995) Low pH induces swiveling of the glycoprotein heterodimers in the Semliki forest virus spike complex. *Cell* **81**, 715–725.

Furcinitti PS, van Oostrum J, Burnett RM. (1989) Adenovirus polypeptide IX revealed as capsid cement by difference images from electron microscopy and crystallography. *EMBO J.* **8**, 3563–3570.

Furrer P, Bednar J, Dubochet J, Hamiche A, Purnell A. (1995) DNA at the entry–exit of the nucleosome observed by cryoelectron microscopy. *J. Struct. Biol.* **114**, 177–183.

Gogol EP, Steven ES, von Hippel PH. (1991) Structure and assembly of the *Escherichia coli* transcription termination factor Rho and its interactions with RNA. I. Cryoelectron microscopic studies. *J. Mol. Biol.* **221**, 1127–1138.

Grief C, Nermut MV, Hockley DJ. (1994) A morphological and immunolabelling study of freeze-substituted human and simian immunodeficiency viruses. *Micron* **25**, 119–128.

Guo XW, Smith PR, Cognon B, D'Arcangelis D, Dolginova E, Mannella CA. (1995) Molecular design of the voltage-dependent, anion-selective channel in the mitochondrial outer membrane. *J. Struct. Biol.* **114**, 41–59.

de Haas F, Traveau J-C, Boisset N, Lambert O, Vinogradov SN, Lamy JN. (1996) Three-dimensional reconstruction of the chlorocruorin of the polychaete annelid *Eudistylia vancouverii. J. Mol. Biol.* **255**, 140–153.

Harris JR. (1988) Electron microscopy of cholesterol. *Micron Microsc. Acta* **19**, 19–32.

Harris JR. (1993) The ultrastructure of multinucleate giant cells. *Micron* **24**, 173–231.

Harris JR, Gebauer W, Söhngen SM, Markl J. (1995) Keyhole limpet hemocyanin (KLH): purification of intact KLH1 through selective dissociation of KLH2. *Micron* **26**, 201–212

ten Heggeler-Bordier B, Wahli W, Adrian M, Stasiak A, Dubochet J. (1992) The apical localization of transcribing RNA polymerases on supercoiled DNA prevents their rotation around the template. *EMBO J.* **11**, 667–772.

Henderson R, Baldwin JM, Ceska TA, Beckmann E, Zemlin F, Downing K. (1990) A model for the structure of bacteriorhodopsin based on high resolution electron cryomicroscopy. *J. Mol. Biol.* **213**, 899–929.

Hewat EA, Booth TF, Roy P. (1992) Structure of bluetongue virus particles by cryoelectron microscopy. *J. Struct. Biol.* **109**, 61–69.

Hewat EA, Booth TF, Roy P. (1994) Structure of correctly self-assembled bluetongue virus-like particles. *J. Struct. Biol.* **112**, 183–191.

Horne RW, Pasquali-Ronchetti I. (1974) A negative staining–carbon film technique for studying viruses in the electron microscope. I. Preparation procedures for examining icosahedral and filamentous viruses. *J. Ultrastruct. Res.* **47**, 361–383.

Jap BK, Li H. (1995) Structure of the osmo-regulated H_2O-channel, AQP-CHIP, in projection at 3.5 Å resolution. *J. Mol. Biol.* **251**, 413–420.

Jontes JD, Wilson-Kubalek EM, Milligan RA. (1995) A 32° tail swing in brush border myosin I on ADP release. *Nature* **378**, 751–753.

Klosgen B, Helfrich W. (1993) Special features of phosphatidylcholine vesicles as seen in cryo-transmission electron microscopy. *Eur. Biophys. J.* **22**, 329–340.

Kühlbrandt W, Wang DN, Fujiyoshi Y. (1994) Atomic model of plant light-harvesting complex determined by electron crystallography. *Nature* **367**, 614–621.

Lambert O, Boisset N, Penczek P, Lamy J, Taveau J-C, Frank J, Lamy JN. (1994) Quaternary structure of *Octopus vulgaris* hemocyanin. Three-dimensional reconstruction of frozen-hydrated specimens and intramolecular location of functional units *O,e* and *O,b. J. Mol. Biol.* **238**, 75–87.

Lambert O, Boisset N, Taveau J-C, Preaux G, Lamy JN. (1995a) Three-dimensional reconstruction of the αD and βC-hemocyanins of *Helix pomatia* from frozen-hydrated specimens. *J. Mol. Biol.* **248**, 431–448.

Lambert O, Taveau J-C, Boisset N, Lamy JN. (1995b) Three-dimensional reconstruction of the hemocyanin of the protobranch bivalve mollusc *Nucula hanleyi* from frozen-hydrated specimens. *Arch. Biochem. Biophys.* **319**, 231–243.

Lepault J. (1985) Cryo-electron microscopy of helical particles of TMV and T4 polyheads. *J. Microsc.* **140**, 73–80.

Mandelkow EM, Mandelkow E, Milligan RA. (1991) Microtubule dynamics and microtubule caps: a time-resolved electron microscopy study. *J. Cell Biol.* **114**, 977–991.

Mandelkow E, Song YH, Schweers O, Marx A, Mandelkow EM. (1995) On the structure of microtubules, tau, and paired helical filaments. *Neurobiol. Aging* **16**, 347–354.

Massover WH. (1986) A comparison of frozen-hydrated unstained and dried negatively-stained specimen preparation for electron microscopy: ferritin and apoferritin. In *Proceedings of the 44th Annual Meeting of the Electron Microscopy Society of America* (ed. GW Bailey). San Francisco Press, San Francisco, pp. 349–350.

Massover WH. (1993) Ultrastructure of ferritin and apoferritin: a review. *Micron* **24**, 389–437.

Massover WH, Adrian M. (1986) Ultrastructure of vitrified (frozen-hydrated) ferritin and apoferritin. In *Proceedings of the 11th International Congress on Electron Microscopy.* Japanese Society for Electron Microscopy, pp. 2427–2428.

McKenna R, Llag LL, Rossmann MG. (1994) Analysis of the single-stranded DNA bacteriophage phi X174, refined at a resolution of 3.0 Å. *J. Mol. Biol.* **237**, 517–543.

Menetret JF, Schröder RR, Hofmann W. (1990) Cryo-electron microscopic studies of relaxed striated muscle thick filaments. *J. Musc. Res. Cell Motil.* **11**, 1–11.

Metcalf P, Cyrklaff M, Adrian M. (1991) The three-dimensional structure of reovirus obtained by cryo-electron microscopy. *EMBO J.* **10**, 3129–3136.

Morgan DG, Owen C, Melanson LA, DeRosier DJ. (1995) Structure of bacterial flagellar filaments at 11 Å resolution: packing of the alpha-helices. *J. Mol. Biol.* **249**, 88–110.

Nermut MV, Grief C, Hashmi S, Hockley DJ. (1993) Further evidence of icosahedral symmetry in human and simian immunodeficiency virus. *AIDS Res. Hum. Retrov.* **9**, 929–938.

Nermut MV, Hockley DJ, Jowett JBM, Jones IM, Garreau M, Thomas D. (1994) Fullerene-like organization of HIV gag-protein shell in virus-like particles produced by recombinant baculovirus. *Virology* **198**, 288–296.

Newcombe WW, Trus BL, Booy FP, Steven AC, Wall JS, Brown JC. (1993) Structure of the herpes simplex virus capsid. Molecular composition of the pentons and the triplexes. *J. Mol. Biol.* **232**, 499–511.

Nogales E, Wolf SG, Zhang SX, Downing KH. (1995a) Preservation of 2-D crystals of tubulin for electron crystallography. *J. Struct. Biol.* **115**, 199–208.

Nogales E, Wolf SG, Khan IA, Luduena RF, Downing KH. (1995b) Structure of tubulin at 6.5 Å and location of the taxol-binding site. *Nature* **375**, 424–427.

O'Toole E, Wray G, Kremer J, McIntosh JR. (1993) High voltage cryomicroscopy of human blood platelets. *J. Struct. Biol.* **110**, 55–66.

Paredes AM, Brown DT, Rothnagel R, Chiu W, Schoepp RJ, Johnston RE, Prasad BVV. (1993) Three-dimensional structure of a membrane-containing virus. *Proc. Natl Acad. Sci. USA* **90**, 9095–9099.

Pum D, Weinhandl M, Hödl C, Sleytr UB. (1993) Large-scale recrystallization of the S-layer of *Bacillus coagulans* E38-66 at the air/water interface and on lipid films. *J. Bacteriol.* **175**, 2762–2766.

Rao Z, Belyaev AS, Fry E, Roy P, Jones IM, Stuart DI. (1995) Crystal structure of SIV matrix antigen and implications for virus assembly. *Nature* **378**, 743–747.

Rossmann MG. (1994) Viral cell recognition and entry. *Protein Sci.* **3**, 1712–1725.

Rossmann MG, Olson NH, Kolatkar PR, Oliveira MA, Cheng RH, Greve JM, McClelland A, Baker TS. (1994) Crystallographic and cryo EM analysis of virion–receptor interactions. *Arch. Virol.* (Suppl.) **9**, 531–541.

Schatz M, Orlova EV, Dube P, Jager J, van Heel M. (1995) Structure of *Lumbricus terrestris* hemoglobin at 30 Å resolution determined using angular reconstruction. *J. Struct. Biol.* **114**, 28–40.

Schröder RR, Manstein DJ, Jahn W, Holden H, Rayment I, Holmes KD, Spudich JA. (1993) Three-dimensional atomic model of F-actin decorated with *Dictyostelium* myosin S1. *Nature* **364**, 171–174.

Shaw AL, Rothnagel R, Chen D, Ramig RF, Chiu W, Prasad BVV. (1993) Three-dimensonal visualization of the rotavirus hemagglutinin structure. *Cell* **74**, 693–701.

Siegel DP, Green WJ, Talmon Y. (1994) The mechanism of lamellar-to-inverted phase transitions: a study using temperature-jump cryo-electron microscopy. *Biophys. J.* **66**, 402–414.

Smith TJ, Cheng RH, Olson NH, Peterson P, Chase E, Kuhn RJ, Baker TS. (1995) Putative receptor binding sites on alphaviruses as visualized by cryoelectron microscopy. *Proc. Natl Acad. Sci. USA* **92**, 10648–10652.

Sosinsky GE, Francis NR, Stallmeyer MB, DeRosier DJ. (1992) Substructure of the flagellar basal body of *Salmonella typhimurium*. *J. Mol. Biol.* **223**, 171–184.

Spin JM, Atkinson D. (1995) Cryoelectron microscopy of low density lipoprotein in vitreous ice. *Biophys. J.* **68**, 2115–2129.

Stewart M. (1991) Transmission electron microscopy of vitrified biological macromolecular assemblies. In *Electron Microscopy in Biology: a Practical Approach,* (ed. JR Harris). IRL Press, Oxford, pp. 229–242.

Stewart PL, Burnett RM. (1995) Adenovirus structure by X-ray crystallography and electron microscopy. *Curr. Top. Microbiol. Immunol.* **199**, 25–38.

Stewart PL, Burnett RM, Cyrklaff M, Fuller SD. (1991) Image reconstruction reveals the complex molecular organization of adenovirus. *Cell* **67**, 145–154.

Stewart PL, Fuller SD, Burnett RM. (1993) Difference imaging of adenovirus: bridging the resolution gap between X-ray crystallography and electron microscopy. *EMBO J.* **12**, 2589–2599.

Stokes DL, Green NM. (1990) Structure of Ca ATPase: electron microscopy of frozen-hydrated crystals at 6 Å resolution in projection. *J. Mol. Biol.* **213**, 529–538.

Stoops JK, Schroeter JP, Bretaudiere JP, Olson NH, Baker TS, Strickland DK. (1991) Structural studies of human α_2-macroglobulin: concordance between projected views obtained by negative-stain and cryoelectron microscopy. *J. Struct. Biol.* **106**, 172–178.

Tahara Y, Fujiyoshi Y. (1994) A new method to measure bilayer thickness: cryoelectron microscopy of frozen hydrated liposomes and image simulation. *Micron* **25**, 141–149.

Unger VM, Schertler GFX. (1995) Low resolution structure of bovine rhodopsin determined by electron cryo-microscopy. *Biophys. J.* **68**, 1776–1786.

Vogel RH, Provencher SW, von Bonsdorff CH, Adrian M, Dubochet J. (1986) Envelope structure of Semliki Forest virus reconstructed from cryo-electron micrographs. *Nature* **320**, 533–535.

Wade RH, Chrétien D. (1993) Cryoelectron microscopy of microtubules *J. Struct. Biol.* **110**, 1–27.

Wagenknecht T, Berkowitz J, Grassucci R, Timerman AP, Fleischer S. (1994) Localization of calmodulin binding sites on the ryanodine receptor from skeletal muscle by electron microscopy. *Biophys. J.* **67**, 2286–2295.

Walker M, Trinick J, White H. (1995) Millisecond time resolution electron cryo-microscopy of the M-ATP transient kinetic state of the acto-myosin ATPase. *Biophys J.* **68**, 87S–91S.

Wang G, Porta C, Chen Z, Baker TS, Johnson JE. (1992) Identification of a Fab interaction footprint site on an icosahedral virus by cryoelectron microscopy and X-ray crystallography. *Nature* **355**, 275–278.

Whittaker M, Wilson-Kubalek EM, Smith JE, Faust L, Milligan RA, Sweeney HL. (1995) A 35 Å movement of smooth muscle myosin on ADP release. *Nature* **378**, 748–751.

Zhou ZH, Prasad BV, Jakana J, Rixon FJ, Chiu W. (1994) Protein subunit structures in the herpes simplex virus A-capsid determined from 400 kV spot-scan electron cryomicroscopy. *J. Mol. Biol.* **242**, 456–469.

9 Future Prospects

A computer search of the recent literature will reveal that both negative staining and cryoelectron microscopy of unstained vitrified specimens continue to be widely used (e.g. Marr *et al.*, 1996; Zhao *et al.*, 1995). The marked decline in the use of negative staining, predicted by some 10 or more years ago, in reality has not happened and the technique continues to be useful and even expand in some areas. We have proposed (Harris and Horne, 1994) that with further technical development negative staining should continue to be valuable for macromolecular studies, and it is now clear that improved resolutions also can be achieved. The increasing availability of electron microscopes with cryospecimen holders means that the study of negatively stained specimens prepared in the presence of carbohydrate (trehalose or glucose) can benefit from cooling to temperatures below −170°C (Orlova *et al.*, unpublished observations). Under these conditions (with minimal adsorption of molecules to the carbon film) the achievable resolution from negative staining can be better than 15 Å within 3-D reconstructions, as shown by Dube *et al.* (1995) for the earthworm, *Lumbricus terrestris,* haemoglobin (*Figure 9.1*). The amplitude contrast is higher within negatively stained specimens than within unstained specimens in vitreous ice, so images can be recorded closer to Scherzer focus and within the plateau of the contrast transfer function, which may ultimately yield resolutions better than 10 Å. Indeed, it should be borne in mind that specimens embedded within a dried layer of negative stain also appear to be considerably less labile within the electron beam than unstained specimens in vitreous ice. However, this may be predominately lower resolution information that is being preserved by the negative stain, because of the inherent limitations of embedding in a high contrast medium. The generally held belief that negative staining will reveal only the surface contour of a protein is undoubtedly correct (Henderson, 1995). Providing the negative staining ion can penetrate the water-space within and around an intricately folded polypeptide chain then meaningful detail of quaternary structure will be obtained. Once the negative stain is unable to enter a very small water-filled surface channel or totally enclosed compartment within a protein, the technique will clearly have come to its limitation. The definition of individual *domains* of a protein can often provide valuable information; this should now be more routinely achievable from negative staining at resolutions in the range 10–15 Å.

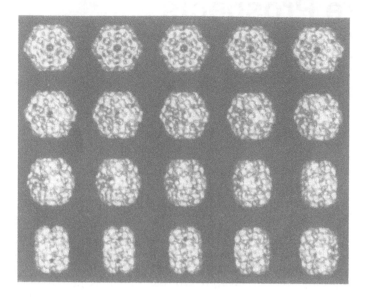

Figure 9.1: 3-D reconstruction of earthworm haemoglobin.
Earthworm haemoglobin: 3-D reconstruction (IMAGIC-5) at 15 Å from a low temperature study of ammonium molydate–glucose specimens (Dube *et al.*, 1995). This series of tilted molecules (a continuous stereo sequence) can be observed as a *magic-eye* stereo-array, with a little practice (cf. *Figure 5.33*). Reconstruction courtesy of Marin van Heel and Prakash Dube.

Because of past problems and the limitations associated with negative staining, some investigators now prefer to use the term 'high contrast embedding techniques(s)' instead of negative staining. It is possible that in time this new terminology may be widely accepted and used, but it is unlikely to replace the long-established and well-understood term 'negative staining' completely, except perhaps by those performing *only* high resolution low temperature TEM studies.

The limitations of negative staining, in terms of the associated problems of partial depth embedding, adsorption deformation, orientation restriction on carbon support films, stain exclusion at the site of adsorption and particle flattening (most pronounced in shallow stain) are well recognized and certainly no longer underestimated. Steps can now be taken to avoid such drawbacks. Direct mixing of sample and negative stain prior to application to carbon support films can create increased particle mobility prior to drying of the stain and the inclusion of carbohydrate (glucose or trehalose) can increase the thickness and protection provided by the embedding film of stain. The removal of a high concentration of salt or other solute from the sample material prior to mixing with the stain cannot always be achieved. In these instances initial adsorption to a carbon support film can be considered beneficial since such salts and solutes can be washed away before negative staining (see Section 3.3). It appears that negative staining with ammonium molybdate (AM) or AM–trehalose (and the other anionic negative stains) induces the release of at least some of the adsorbed specimen material into the fluid film prior to drying, and may thereby prevent or reduce the restriction of particle orientation due to selective adsorption. Uranyl acetate does not appear to act in this way, probably because of its cationic nature and acidity; in addition, it should be noted that it is not desirable to mix uranyl acetate with a biological sample prior to application to the grid, since pronounced precipitation or aggregation may occur.

The production of negatively stained specimens on holey carbon support films, with sample and stain unsupported across the holes, has not been widely considered as a useful approach in the past. However, this possibility was utilized by Fernández-Morán et al. (1968) and Mellema and Klug (1972), and attempts have been made to support material in negative stain alone across the corners of fine mesh EM grids (Malech and Albert, 1979). In these instances the instability and mobility of the thin layer of stain in the electron beam often prevented satisfactory image recording. However, with the introduction of trehalose as a constituent of the negative stain mixture and the known property of trehalose to create vitreous protective layers (Green and Angel, 1989; Sikora et al., 1994) this situation could be considerably improved, particularly when trehalose-containing negative stains are used in conjunction with holey carbon support films. That a thin frozen-hydrated film of aqueous trehalose can remain as a layer across a hole following drying has already been shown (*Figure 7.5*). Extending this approach to negative staining indicates that satisfactory image recording may now be achieved for unsupported material across the holes of holey carbon support films, as demonstrated by *Figures 9.2* and *9.3*. As with specimens of unstained frozen-hydrated material, the liposome particles and KLH1 didecamers shown here have a tendency to move to the edges of holes in the support film where the fluid film is thicker prior to drying, indicating that surface forces at the air–fluid interface influence the sample material and are likely to impose some element of selective orientation. Thus, although the limitations of carbon-induced particle adsorption have been removed, the influence of surface tension forces, evaporation and drying effects cannot be completely avoided, even if such a negatively stained sample is rapidly frozen before drying of the fluid film (see below).

The production of 2-D crystals directly upon carbon and holey carbon support films has been addressed by several investigators, with some success (Holzenburg, 1988; Keegstra and van Bruggen, 1980). This approach seems worthy of further investigation, either for the production of negatively stained or unstained frozen-hydrated specimens. If the adsorption properties of the carbon can be blocked (e.g. with octyl glucoside or polyethylene glycol) then the molecular freedom needed to form the precise intermolecular contacts necessary for crystallization may come into play, creating a more satisfactory crystallization system, somewhat equivalent to the negative staining–carbon film procedure on the hydrophilic surface

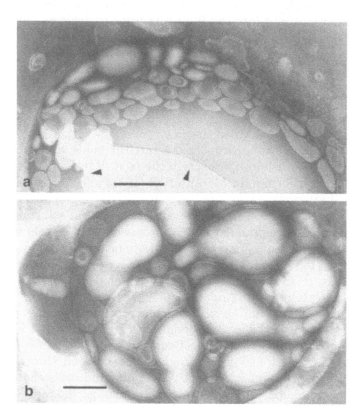

Figure 9.2: Clustering of phosphatidylcholine–cholesterol liposome particles.
Phosphatidylcholine–cholesterol liposome particles suspended in 2.5% AM–0.5% trehalose across a hole in a carbon support film. Note the clustering of the particles to the edge of the hole, with breakage of the thin AM–trehalose film (arrows, a) (cf. negatively stained liposomes supported on a carbon film, *Figure 5.10*). In (b) a smaller hole is completely filled with liposomes and stain. The scale bars indicate 200 nm (a) and 100 nm (b).

of mica (see Section 3.5). It is possible that 2-D crystal formation occurs primarily at the air–fluid interface (Harris and Holzenburg, 1995; Pum *et al.*, 1993), in which case an *unreactive* carbon surface should be a suitable substrate. With holey carbon support films, crystallization salts, such as ammonium sulphate and AM, and other agents such as PEG may be removed by appropriate water or buffer washing from one side of the grid (e.g. immediately before or even after complete drying of the AM–PEG), leaving protein or virus particles and 2-D crystals 'trapped' at the opposite air–fluid interface spanning the holes (Cyrklaff *et al.*, 1994; Dubochet *et al.*, 1985, 1986) prior to the addition of negative stain or rapid freezing.

Frozen specimens can be prepared directly from thin aqueous films of biological material suspended in a higher concentration of negative stain

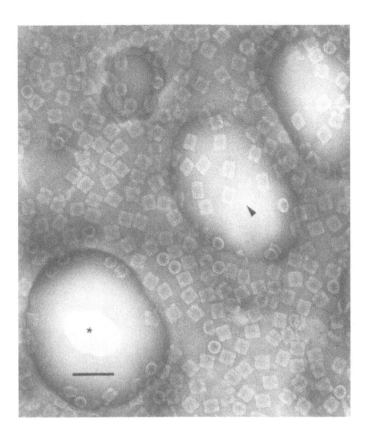

Figure 9.3: KLH1 didecamers spread across a glow-discharge treated holey carbon film.
Purified KLH1 didecamers spread across a glow-discharge treated holey carbon film after mixing with an equal volume of 5% AM–1% trehalose (pH 7.0). Note the excellent contrast of the molecules (arrow) and the thinning and breakage of the AM–trehalose film (star). The scale bar indicates 100 nm.

solution (e.g. 20% AM, pH 7.0); such investigations are at a very early stage (see *Frontispiece, Figures 9.4* and *9.5*). This high contrast rapid freezing approach may include the benefits of vitrification along with the increased contrast implicit within negatively stained images, but investigators need to be aware of the possible damage to biological materials resulting from suspension in higher concentrations of negative stains. These are necessary because of the very thin aqueous film produced prior to freezing, compared to the slightly thicker aqueous film used in conventional negative staining. Such conditions could, however, offer considerable advantages for the combined electron microscopical and biochemical study of dissociating and reassociating macromolecular systems. This type of approach has recently been followed by Boettcher *et al.* (1996), who

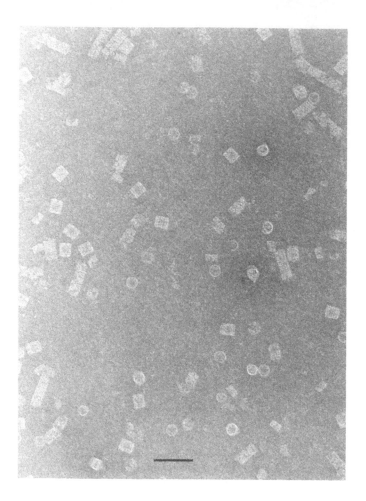

Figure 9.4: Frozen-hydrated KLH1 in the presence of AM.

Frozen-hydrated KLH1 following reassociation from the purified subunit, with the specimen prepared directly from thin aqueous film containing 16% AM (pH 7.0). Decamers, didecamers and short tubular polymers are present. Note the excellent contrast and preservation of the molecules. The scale bar indicates 100 nm. Micrograph courtesy of Marc Adrian. (Cf. *Figures 5.29* and *9.3* showing conventionally negatively stained KLH and *Figure 8.8* of frozen-hydrated unstained KLH.)

used a low concentration of phosphotungstic acid (0.5–2.0%) to stabilize a helical/twisted ribbon aqueous self-organization of n-octyl-D-gluconamide and then directly created vitreous ice films, still in the presence of the negative stain.

In general, the number of applications of cryoelectron microscopy of unstained frozen-hydrated biological specimens has steadily increased in the past decade. Some investigators pursue only this strict 'unstained cryo' approach, although many perform initial negative staining investigations with the hope of subsequently obtaining superior data from unstained

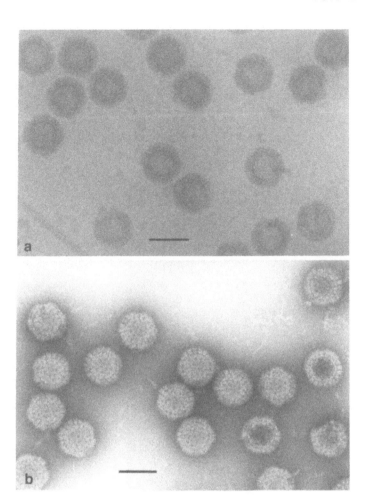

Figure 9.5: Vitrified rotavirus.

Cryoelectron micrographs of rotavirus imaged (a) unstained in vitreous ice, and (b) in vitreous 16% AM. Note the close agreement between the two images, the excellent preservation of the capsid detail in both cases and the definition of the viral core. The scale bars indicate 100 nm. Micrographs courtesy of Marc Adrian.

frozen hydrated specimens, or indeed cryonegative staining (Dube *et al.*, 1995). The two technical approaches are, therefore, interactive largely complementary and mutually supportive.

It has been predicted by David DeRosier (1993) that by the turn of the century further improvement from unstained specimens will be achieved and enable a resolution of 9 Å for icosahedral viruses, 4 Å for helical structures and even 2.8 Å for proteins within 2-D crystals. At these resolutions it should be possible to locate the α-helices within viral subunits, the peptide chain within the subunits of helical proteins and the atoms of individual

Figure 9.6:   A cryoelectron micrograph of tobacco mosaic virus.

A cryoelectron micrograph from a specimen of tobacco mosaic virus (TMV) rapidly frozen in the presence of approximately 4.5% trehalose. This low electron dose image shows several viruses, with the 2.3 nm linear repeat showing directly. Inset: the computed Fourier transform/ power spectrum from a single virus, with a strong 2.3 nm reflection (arrows) (see Schröder *et al.*, 1990; Smith and Langmore, 1992). The scale bar indicates 100 nm. Cryoelectron micrograph, courtesy of Marc Adrian.

amino acids within proteins in 2-D crystals. Most significantly, studies on an increasing number of membrane proteins, which are not readily available as 3-D crystals for X-ray diffraction analysis, are likely to make steady progress (Dorset, 1995; Kühlbrandt, 1992).

Such an increase in overall resolution is likely to be achieved through a combination of technical improvements at the specimen level, at the electron microscope and within the various software packages used for image processing and 3-D reconstruction (see Chapter 10). For instance the in clusion of trehalose in unstained samples before freezing could directly provide some increased sample protection within the electron beam (Marc Adrian, personal communication); an example of unstained TMV rapidly frozen in the presence of 4.5% trehalose, is shown in *Figure 9.6*. An increased appreciation of the benefits to be gained from the study of a very large number of single particles (viruses or protein molecules) in order to have a truly homogeneous data set containing a maximal number of possible image orientations would appear to be important. The more widespread handling of large data sets will undoubtedly require improvement in the software, most of which is currently far from being *user friendly*. The production of images from more perfectly polymerized undistorted helical structures and highly ordered 2-D crystals will undoubtedly be of assistance. Thus, steady improvement in specimen preparation and the *quality* of the biological samples are areas that should not be neglected.

The available software is designed to provide corrections for such defects, although there is some loss of absolute resolution compared to that obtained from more perfect specimen material. For both crystallographic

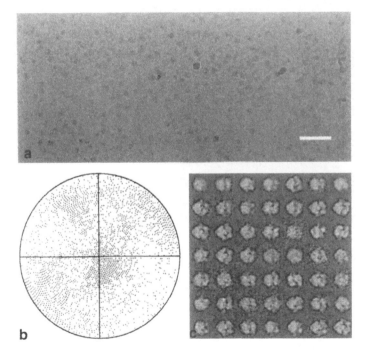

Figure 9.7: Energy filtered images of vitreous water-embedded *E. coli* ribosomes.
Energy filtered images of vitreous water-embedded *E. coli* ribosomes (a), the predominant orientations of 4300 selected particles (b) and averages of clusters found by classification (c). For futher details see Frank *et al.* (1995). The scale bar (a) indicates 100 nm. Figure courtesy of Joachim Frank. Reprinted with permission from Frank *et al.* (1995) A model of protein synthesis based on cryo-electron microscopy of the *E. coli* ribosome. *Nature* 376, 441–444. Copyright 1995 Macmillan Magazines Ltd.

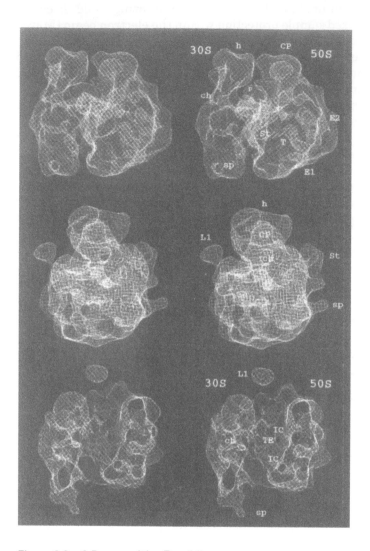

Figure 9.8: 3-D maps of the *E. coli* ribosome.

3-D maps (stereo pairs) of the *E. coli* ribosome at 25 Å, presented from different viewing angles. For further details see Frank *et al.* (1995). The image reconstruction shown here was obtained using the SPIDER software package utilizing the simultaneous iterative reconstruction technique. Correction for contrast transfer function (CTF) was included; this restored the correct weight for the low spatial frequencies. Figure courtesy of Joachim Frank. Reprinted with permission from Frank *et al.* (1995) A model of protein synthesis based on cryo-electron microscopy of the *E. coli* ribosome. *Nature* 376, 441–444. Copyright 1995 Macmillan Magazines Ltd. For full information see this publication.

and single particle studies, the spot-scan technique has been introduced to reduce the problem of specimen movement during image recording (Downing, 1991); see also Chapter 10. Automated tomography is also likely to be of some benefit for the controlled production of tilt series from single particles and 2-D crystals (Dirksen *et al.*, 1995; Fung *et al.*, 1996). The availability of zero loss energy filtration on some cryotransmission electron microscopes has enabled an increased image contrast to be obtained from thin unstained specimens in vitreous ice (Schröder, 1992; Schröder *et al.*, 1990; Smith and Langmore, 1992). This can make the alignment of particles more accurate and thereby improve resolution. The improved contrast also makes it possible to perform contrast transfer function corrections with a reduced danger of introducing artefacts from image components that do not follow elastic theory (Joachim Frank, personal communication). The current resolution of 24 Å reconstruction achieved for the *E. coli* 70S ribosome in vitreous ice using energy filtering and contrast transfer function correction must be considered to be a considerable achievement (Frank *et al.*, 1995), as shown in *Figures 9.7* and *9.8*. Further steady improvement is likely in the near future, for the ribosome and many other biological assemblies.

In nearly all cases it must be emphasized that the study of structure alone is rarely considered in isolation from functional studies. For instance, the aforementioned study on the 70S ribosome is firmly linked to the understanding of protein synthesis. In the case of viral structure, viral biosynthesis, the understanding of the fusion of viruses with cellular membranes (Fuller *et al.*, 1995) and the interaction of neutralizing or non-neutralizing antibodies (Mateu, 1995) are all of importance. With the giant extracellular respiratory proteins, conformational change upon oxygen binding and subunit cooperativity in oxygen uptake and release are aspects that are likely to be further pursued by electron microscopical studies, as is the role of the ubiquitous chaperonins in protein folding. Thus, the study of structure and function must and will continue to go hand-in-hand.

References

Boettcher C, Stark H, van Heel M. (1996) Stacked bilayer helices: a new structural organization of amphiphilic molecules. *Ultramicroscopy,* in press.

Cyrklaff M, Roos N, Gross H, Dubochet J. (1994) Particle-surface interaction in thin vitrified films for cryo-electron microscopy. *J. Microsc.* **175**, 135–142.

DeRosier DJ. (1993) Turn-of-the-century electron microscopy. *Curr. Biol.* **3**, 690–692.

Dirksen K, Typke D, Hegerl R, Walz J, Sackman E, Baumeister W. (1995) Three-dimensional structure of lipid vesicles embedded in vitreous ice and investigated by automated electron tomography. *Biophys J.* **68**, 1416–1422.

Dorset DL. (1995) Direct structure determination by electron crystallography: protein data sets. *Micron* **26**, 511–520.

Downing KH. (1991) Spot-scan imaging in transmission electron microscopy. *Science* **251**, 53–59.

Dube P, Stark H, Orlova EV, Schatz M, Beckmann E, Zemlin F, van Heel M. (1995) 3D structure of single macromolecules at 15 Å resolution by cryo-microscopy and angular reconstitution. In *JMSA Proceedings, Microscopy and Microanalysis* (eds GW Bailey, MH Ellisman, RA Hennigar, NJ Zaluzec). Jones and Begell, New York, pp. 838–839.

Dubochet J, Adrian M, Lepault J, McDowall AW. (1985) Cryo-electron microscopy of vitrified specimens. *Trends Biol. Sci.* **10**, 143–146.

Dubochet J, Adrian M, Chang J-J, Homo J-C, Lepault J, McDowall AW, Schultz P. (1986) Cryo-electron microscopy of vitrified specimens. *Quart. Rev. Biophys.* **21**, 129–228.

Fernández-Morán H, Marchalonis JJ, Edelman GM. (1968) Electron microscopy of a hemagglutinin from *Limulus polyphemus. J. Mol. Biol.* **32**, 467–469.

Frank J, Zhu J, Penczek P, Li Y, Srivastava S, Verschoor A, Radermacher M, Grassucci R, Lata RK, Agrawal RK. (1995) A model of protein synthesis based on cryo-electron microscopy of the *E. coli* ribosome. *Nature* **376**, 441–444.

Fuller SD, Berriman JA, Butcher SJ, Gowen BE. (1995) Low pH induces swivelling of the glycoprotein heterodimers in the Semliki Forest virus spike complex. *Cell* **81**, 715–725.

Fung JC, Liu W, de Ruijter WJ, Chen H, Abbey CK, Sedat JW, Agard DA. (1996) Toward fully automated high resolution electron tomography. *J. Struct. Biol.* **116**, 181–189.

Green JL, Angell CA. (1989) Phase relations and vitrification in saccharide–water solutions and the trehalose anomaly. *J. Phys. Chem.* **93**, 2880–2882.

Harris JR, Holzenburg A. (1995) Human erythrocyte catalase: 2-D crystal nucleation and production of multiple crystal forms. *J. Struct. Biol.* **115**, 102–112.

Harris JR, Horne RH. (1994) Negative staining: a brief assessment of current technical benefits, limitations and future possibilities. *Micron* **25**, 5–13.

Henderson R. (1995) The potential and limitations of neutrons, electrons and X-rays for atomic resolution microscopy of unstained biological molecules. *Quart. Rev. Biophys.* **28**, 171–193.

Holzenburg A. (1988) Preparation of two-dimensional arrays of soluble proteins as demonstrated for bacterial D-ribulose-1,5-bisphosphate carboxylase/oxygenase. In *Methods in Microbiology* Vol. 20. Academic Press, New York, pp. 341–356.

Keegstra W, van Bruggen EFJ. (1980). A simple way of making a 2-D array. In *Electron Microscopy at Molecular Dimensions* (ed. W Baumeister). Springer-Verlag, Berlin, pp. 318–327.

Kühlbrandt W. (1992) Two-dimensional crystallization of membrane proteins. *Quart. J. Biophys.* **25**, 1–49.

Malech HL, Albert JP. (1979) Negative staining of protein macromolecules: a simple rapid method. *J. Ultrastruct. Res.* **69**, 191–195.

Marr KM, Mastronarde DN, Lyon MK. (1996) Two-dimensional crystals of photosystem II: biochemical characterization, cryoelectron microscopy and localization of the D1 and cytochrome b559 polypeptides. *J. Cell Biol.* **132**, 823–833.

Mateu MG. (1995) Antibody recognition of picornaviruses and escape from neutralization: a structural view. *Virus Res.* **38**, 1–24.

Mellema JE, Klug A. (1972) Quaternary structure of gastropod haemocyanin. *Nature* **239**, 146–150.

Pum W, Weinhandl M, Hödl C, Sleytr UB. (1993) Large-scale recrystallization of the S-layer of *Bacillus coagulans* E38-66 at the air/water interface and on lipid films. *J. Bacteriol.* **175**, 2762–2766.

Schröder RR. (1992) Zero-loss energy-filtered imaging of frozen-hydrated proteins: model calculations and implications for future developments. *J. Microsc.* **166**, 389–400.

Schröder RR, Hofmann W, Memetret J-F. (1990) Zero-loss energy filtering as improved imaging mode in cryoelectron microscopy of frozen-hydrated specimens. *J. Struct. Biol.* **105**, 28–34.

Sikora S, Little AS, Dewey TG. (1994) Room temperature trapping of rhodopsin photointermediates. *Biochemistry* **33**, 4454–4459.

Smith MF, Langmore JP. (1992) Quantitation of molecular densities by cryo-electron microscopy: determination of the radial density distribution of tobacco mosaic virus. *J. Mol. Biol.* **226**, 763–774.

Zhao XZ, Fox JM, Olson NH, Baker TS, Young MJ. (1995) *In vitro* assembly of cowpea chlorotic mottle virus from coat protein expressed in *Escherichia coli* and *in vitro*-transcribed viral cDNA. *Virology* **207**, 486–494.

10 Computer Processing for 2-D and 3-D Image Reconstruction

In this chapter a brief introduction will be given to the available approaches and software for the digital image processing of electron micrographs containing negatively stained and unstained cryoimages and the subsequent 2-D and 3-D reconstruction of averaged images. It must be emphasized that the electron microscopist now needs to become a competent handler of one or more software packages, if not actually a programmer. In general, the available software is not user friendly, often having been produced by skilled programmers for specific purposes over a period of many years; there are a few exceptions (even PC-based systems running under Windows) and these are likely to increase in number as more people become involved in image processing from electron micrographs and on-line electron images. The simplest approach initially is to define which area of image processing one wishes to pursue (i.e. single particle, crystallographic or helical analysis) and then visit a laboratory where similar work is already in progress. Assistance may be given by direct collaboration or tuition may be given in the use of the available software; the latter could ultimately mean a commitment to many months of work and large expenditure, such as a PC controlled high resolution scanning densitometer or scanning CCD camera system, a high resolution B/W printer or a colour videoprinter. It is apparent that the whole field is currently in the process of slowly moving from the few specialist laboratories with extensive computer facilities, back-up and knowledge, to a situation of more general availability, a trend that will surely continue. Most laboratories have retained their steadily developing software within the public domain, whereas a few others have moved into the commercial area for the further establishment and marketing of their image processing software (see *Table 10.1* and Appendix B).

Several books have been devoted to or include sections on image processing. The first, *Image Analysis, Enhancement and Interpretation* by Derek L. Misell (North-Holland, 1978) remains useful, as is the chapter 'Image analysis of electron micrographs' by Mike F. Moody, in the book *Biophysical Electron Microscopy* (P.W. Hawkes and U. Valdré, Academic Press, 1990) and supplement 6 of *Scanning Microscopy*, 'Signal and image processing in microscopy and microanalysis' (Scanning Microscopy International, 1992). A recent important addition has been *Three-Dimensional Electron Microscopy of Macromolecular Assemblies* by Joachim Frank (Academic

Table 10.1: Some software available for image processing and 3-D reconstruction

A. Commercial [a]

CRISP (Trimerge, Triview) and ELD	Particularly suitable for analysis of 2-D crystals and electron diffraction patterns.
IMAGIC-5	A diverse package, particularly suitable for single particle analysis.
SEMPER	A diverse package, particularly suitable for 2-D crystal analysis.

B. Public domain/academic [b]

MRC image processing programs	Suitable for diverse applications.
SPIDER and WEB	Particularly suitable for single particle analysis.
The PIC System-III	Suitable for viral structure.
SUPRIM	Suitable for single particle analysis.
Extensible and Object-orientated System (EOS)	Particularly suitable for macromolecules.
The Integrated Crystallographic Environment (ICE)	Particularly suitable for crystallographic analysis.
The Brandeis Helical Package	Helical analysis.
Image Visualization Environment (IVE)	Diverse applications.
PHOELIX	Helical image processing.
XMIPP	Particularly suitable for single particle image processing.
PTOOL	Useful for spot-scan images.
AVS	For 3-D visualization of molecules.
IMOD	For visualization of 3-D image data.
SIGMA	Software for imagery and graphics for molecular architecture.

[a] Company addresses given in Appendix B.
[b] Taken from *J. Struct. Biol.* 116, No. 1. (1996), Special Issue: Advances in Computational Image Processing for Microscopy. For up-to-date information on available software access http://rcr-www.med.nyu.edu/3dem/homepage.html.

Press, 1996). In addition, a special issue of the *Journal of Structural Biology* has been devoted to 'Advances in computational image processing for microscopy' (Vol. 116, No. 1, 1996). This issue contains a very useful collection of articles, that are likely to be of lasting value. An introductory overview from Carracher and Smith (1996) sets the scene by providing some historical background and an assessment of the present state of the art in image processing. Due recognition must be given to the early contribution from Aaron Klug and his colleagues at the MRC Laboratory of Molecular Biology in Cambridge, in particular David DeRosier, Tony Crowther, Linda Amos, Richard Henderson and Nigel Unwin. The strong X-ray crystallographic roots within this laboratory were extended to a consideration of the electron microscopy of thin 3-D and 2-D protein crystals, helical assemblies and icosahedral viruses, and laid the foundations for much of present-day image processing. The use of Fourier synthesis remains the fundamental strength of this approach (Crowther, 1971; DeRosier and Moore, 1970; Mellema and Klug, 1972). Definition of the rotational symmetry of negatively stained structures by computer rotational harmonic analysis rather than by simple photographic rotational enhancement was also an important contribution (Crowther and Amos, 1971). This area of image analysis continues to be of significance, particularly for the detection of the rotational symmetries of macromolecules in noisy electron micrographs (Kocsis *et al.*, 1995).

Independently, Walter Hoppe and his colleagues in Munich, in particular Reiner Hegerl, pursued single molecule analysis, an area that has subsequently greatly expanded within the hands of Marin van Heel, Joachim Frank and their colleagues (van Heel *et al.*, 1992; Penczek *et al.*, 1992). An appreciation of the importance of the defocus aberrations due to the oscillating frequencies of the contrast transfer function by Erickson and Klug (1971) has had a major impact throughout the whole of biomolecular image processing (see Zhou *et al.*, 1996). The ability to 'straighten or unbend' 2-D crystals and helical filamentous structures, merge data and perform cross-correlation corrections have also played important roles (Saxton, 1996).

A word of caution is perhaps appropriate, before progressing to a consideration of some specific areas, since a few extremely unsatisfactory situations and errors have appeared in the literature that do no justice to the field of macromolecular electron microscopy. Image processing should perhaps only be applied in situations where the raw data are of sufficient quality to justify the time that will be required; direct visual interpretations should indicate that image processing is likely to provide a *valid* and significant addition to the existing literature. For instance, if it is clear that the material under investigation is unstable, perhaps due to partial degradation or dissociation, or to staining and preparation artefacts, considerable care must be taken. Image processing should never be used in an attempt to generate improved data from essentially unsatisfactory and perhaps excessively variable raw images. Although generally of great value in image processing, the imposition of particle symmetry is also an area where some care must be taken (Harris *et al.*, 1992), particularly when extending interpretations through image simulation packages (see Lamy *et al.*, 1993).

10.1 Single particle analysis

It is encouraging to note that the available software packages for image processing of single molecules exhibit considerable inter-reliability. Thus, independent 3-D reconstructions from IMAGIC-5 and SPIDER/WEB of the skeletal muscle calcium release channel/ryanodine receptor from unstained molecules in vitreous ice exhibit an almost identical structure (Serysheva *et al.*, 1995; Wagenknecht *et al.*, 1995). Currently, the resolution of most macromolecular 3-D reconstructions from vitreous ice is in the order of

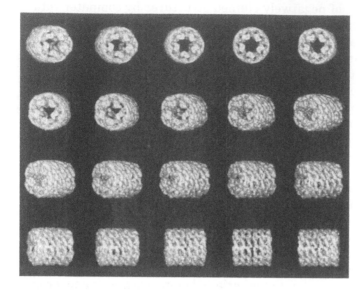

Figure 10.1: 3-D reconstruction of KLH1 didecamer at 15 Å.

A stereo array of progressively tilted KLH1 didecamers (a continuous stereo sequence), following 3-D image processing and reconstruction within the IMAGIC-5 software system. The original low dose micrographs were produced from specimens negatively stained with 2% ammonium molybdate containing 1% glucose, and studied at 4 K (Orlova *et al.*, unpublished observations). The resolution achieved is better than 15 Å, indicating the considerable possibilities for improved low temperature negative staining (i.e. embedding in high contrast media) when combined with the powerful possibilites of single particle image processing. Photograph courtesy of Marin van Heel, Prakash Dube and Elena Orlova.

Figure 10.2: The A-capsid of unstained herpes simplex virus type 1 in vitreous ice.

A small area from a 400 kV spot-scan electron micrograph of the A-capsid of unstained herpes simplex virus type 1, embedded in vitreous ice, at 3.1 μm defocus. The scale bar indicates 300 nm. Micrograph courtesy of Wah Chiu. Reprinted from Zhou *et al.* (1994) Protein subunit structures in the herpes simplex virus A-capsid determined from 400 kV spot-scan electron cryomicroscopy. *J. Mol. Biol.* 242, 456–469, with permission from Academic Press Ltd.

25–45 Å. This is likely to be improved in the foreseeable future, with the construction of larger data sets and the more routine inclusion of CTF corrections. In the interim, it is possible that 2-D and 3-D image reconstructions from macromolecules negatively stained in the presence of protective carbohydrate and studied at low temperature will routinely result in resolutions of 15 Å and it has been predicted that considerably better than 10 Å will soon be achieved. *Figure 10.1* shows a gallery of tilted 3-D reconstructions at approximately 15 Å, of the KLH1 didecamer, imaged from 2% AM–1% glucose (cf. the reconstruction from unstained molecules in vitreous ice, *Figure 8.9*). Full details of this study will soon be published, together with a detailed appraisal of 3-D reconstruction from molecules freely and totally embedded in high contrast media (negative stain) and an assessment of the advantages of the IMAGIC software system for the recovery of 3-D information from single exposures, rather than a pair of negatives needed by most other systems (Marin van Heel, personal communication). For high molecular mass macromolecules and macromolecular assemblies, single particle image processing is likely to continue to be the predominant line for future 3-D analysis, since the formation and study of 2-D crystals from such material presents its own specific problems once there is extensive overlapping of subunits within the electron images, as occurs extensively in oligomeric proteins (see below).

As indicated in Chapter 8, the single particle analysis of icosahedral viral particles has contributed to a major advance in the understanding of viral structure. Spot–scan (400 kV) images of unstained frozen-hydrated herpes simplex virus A-capsid are shown in *Figure 10.2*, from which Zhou *et al.* (1994) were able to define the protein subunit structure within the capsid, as indicated by the 24 Å reconstruction shown in *Figure 10.3*. Three-dimensional reconstructions of adenovirus have also been important in understanding this group of viruses and considerable integration of X-ray data with that from electron microscopy has been useful (Stewart and Burnett, 1995; Stewart *et al.*, 1991, 1993), as shown in *Figure 10.4* (see also Sections 5.3 and 8.4).

10.2 Crystallographic analysis

Following the ground-breaking electron crystallographic study of Unwin and Henderson (1975) on purple membrane, this approach to 3-D reconstruction from 2-D crystals of membrane and soluble proteins has made steady progress at the highest available resolution (approx. 3 Å) as well as at a more intermediate level (approx. 20 Å). The crystallographic software developed by the MRC-LMB in Cambridge (Crowther *et al.*, 1996) has also been integrated into a number of the available systems and continues to be increasingly used. Thus, most of the available image processing packages offer the facility for 2-D crystallographic analysis with merging of tilt

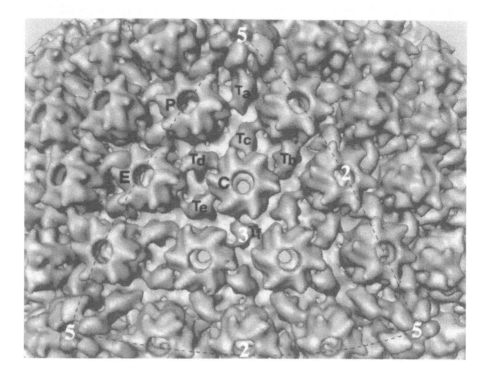

Figure 10.3: Portion of a 3-D map of the A-capsid of herpes simplex virus type 1.

A portion of the 26 Å 3-D map of the A-capsid of herpes simplex virus type 1, as viewed along the icosahedral three-fold axis. One of the 20 triangular faces is outlined, and the isosahedral symmetry axes (5,3,2) are labelled. Reconstruction courtesy of Wah Chiu. Reproduced from Zhou *et al.* (1994) Protein subunit structures in the herpes simplex virus A-capsid determined from 400 kV spot-scan electron cryomicroscopy. *J. Mol. Biol.* 242, 456–469, with permission from Academic Press. For full information see this publication.

series and 3-D reconstruction, including the PC-based system CRISP (Hovmöller, 1992) which has an immediate benefit for the novice of being *user friendly*, although possibly lacking some of the sophistications of other packages. This package also possesses considerable potential for the atomic resolution electron crystallographic analysis of inorganic and organic materials (Weirich *et al.*, 1996)

Two-dimensional crystalline sheets of tubulin can be created during the early stages of polymerization of α and β tubulin into microtubules. Hoenger *et al.* (1995) studied such crystals before and after decoration with the non-claret disjunctional (ncd) motor protein by negative staining with uranyl acetate (*Figure 10.5*), utilizing image processing within the SPECTRA system. By comparison of the two merged tilt series they produced 3-D maps of the tubulin sheets which indicated that the ncd motor protein binds to the crest of a single protofilament, with extensive contacts to both the α and β tubulin monomers and that binding of the motor protein results in a significant conformational change within both tubulin monomers. *Figure 10.6* shows projections of the 3-D maps at approximately 20 Å resolution, describing these structural features. The success of 3-D reconstruction from negatively stained 2-D crystals is dependent upon complete embedding of the protein within negative stain. Whilst this can be achieved for small proteins, with larger oligomeric proteins such as the 16S soluble Mg–ATPase complex, the 20S proteasome and the chaperonin GroEL, partial depth embedding has often precluded the production of *meaningful* 3-D reconstructions. By ensuring complete embedding in negative stain, as is the case in the presence of glucose or trehalose which usually increase the stain depth, it is possible that such problems may be

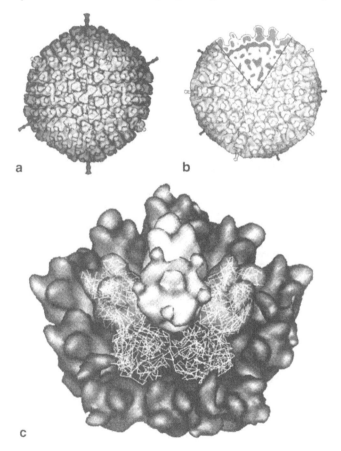

Figure 10.4: 3-D reconstruction of unstained adenovirus type 2 in vitreous ice.
    (a) Viewed along the icosahedral two-fold axis, and (b) along the icosahedral five-fold axis with a sector showing a central slice through the reconstruction. (c) The adenovirus penton surrounded by five hexons, with fitting of the crystallographic peptide backbone of the hexon trimer. Reconstructions courtesy of Steve Fuller. Modified from Stewart *et al.* (1991, 1993).

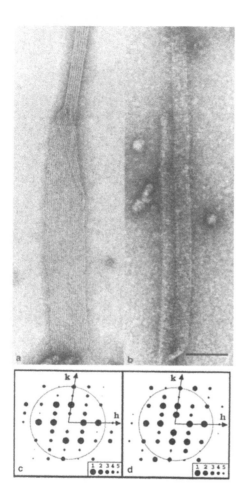

Figure 10.5: 2-D crystalline sheets of tubulin before and after decoration with ncd.

Negatively stained images (2% uranyl acetate) of tubulin sheets (a) and sheets decorated with the motor domain of ncd (non-claret disjunctional protein), a member of the kinesin superfamily (b). (c and d) Show the respective representations of the diffraction data obtained from corresponding images, after correction of the lattice for distortions. The scale bar indicates 100 nm. Figure courtesy of Andreas Hoenger. Reprinted with permission from Hoenger *et al.* (1995) Three-dimensional structure of tubulin–motor-protein complex. *Nature* 376, 271–274. Copyright 1995 Macmillan Magazines Ltd.

largely overcome in the future, even for 2-D crystals of high molecular mass proteins.

As shown in *Figure 10.7a*, 2-D crystals can be produced from human erythrocyte catalase (Harris and Holzenburg, 1995). This enzyme has a molecular mass of 256 kDa, and contains four identical subunits. Randomly positioned single molecules of catalase can be seen at the top of this electron micrograph. Two-dimensional image processing within the CRISP system indicates that the lattice possesses p2 plane group symmetry, with

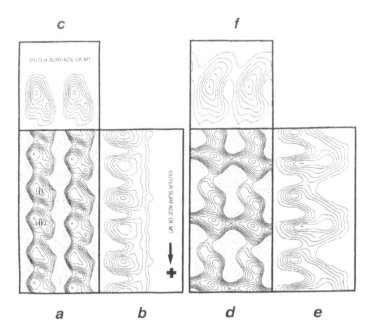

Figure 10.6: Projections of the 3-D maps of tubulin sheets and ncd-decorated sheets.
Projections of the 3-D maps of tubulin sheets (a–c) and ncd-decorated sheets (d–f).
Microtubule polarity is indicated by +. (a) and (d) Show the x–y projections of two protofilaments
(running vertically) in the 3-D map, equivalent to viewing the wall of a microtubule from the
outside. (b) and (e) Show the summation of the density associated with a single protofilament in
the direction of the three-start helix of a microtubule. Figure courtesy of Andreas Hoenger.
Reprinted with permission from Hoenger *et al.* (1995) Three-dimensional structure of tubulin–
motor-protein complex. *Nature* 376, 271–274. Copyright 1995 Macmillan Magazines Ltd.

two molecules per unit cell (*Figure 10.7b*). From this 2-D crystal the averaged projection image suggests that the molecule does not contain four
subunits; this is simply because of partial superimposition of subunits
within the single molecular orientation present in the 2-D crystal, resulting in an equally specific projection image. Catalase is able to generate
several different 2-D crystal forms (Harris and Holzenburg, 1989; Harris
et al., 1993), each with the catalase molecule in a different planar orientation. This enzyme thereby provides the possibility of obtaining some insight into 2-D crystal nucleation and the sites on the molecular surface
that are responsible for intermolecular contact and crystal growth. That
partial depth negative staining does occur was clearly described in our
earlier publications, the difficulty being that although the regions of deeper
stain can be readily defined, complete embedding in negative stain (then
in the absence of carbohydrate) could not always be ensured.

GroEL and other cpn60 chaperonins have been successfully studied by
single particle 2-D projection averaging from negatively stained specimens
(Langer *et al.*, 1992), 3-D averaging from unstained specimens in vitreous

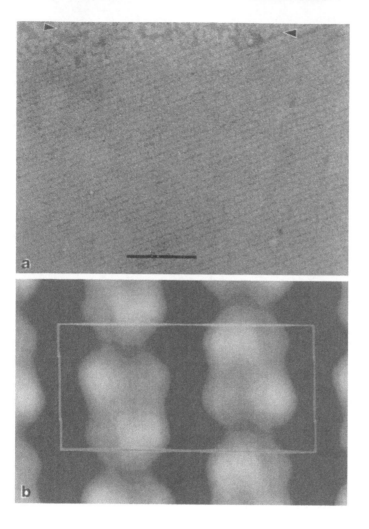

Figure 10.7: Projection image reconstruction from a 2-D crystal of human erythrocyte catalase. A 2-D crystal of human erythrocyte catalase produced by the NS–CF technique (a) with the corresponding p2 crystallographic 2-D image reconstruction produced within CRISP(b), with two 180° rotated molecules in the unit cell. Note the presence of randomly orientated free molecules, arrows (a). The scale bar (a) indicates 100 nm. Modified from Harris and Holzenburg (1995).

ice (Chen *et al.*, 1994) and crystallographic analysis (Harris *et al.*, 1993, 1995; Zahn *et al.*, 1993) using the SEMPER and CRISP systems. *Figure 10.8a* shows small 2-D crystals of GroEL, negatively stained with uranyl acetate, with the cylindrical molecules orientated on their sides. The projection average from a suitably ordered region is given in *Figure 10.8b,* from which it can be seen that the molecules are organized in an extremely unusual manner within the p2 lattice, to accommodate the heptameric rotational symmetry of the molecule. Each molecule is a dimer of two rings

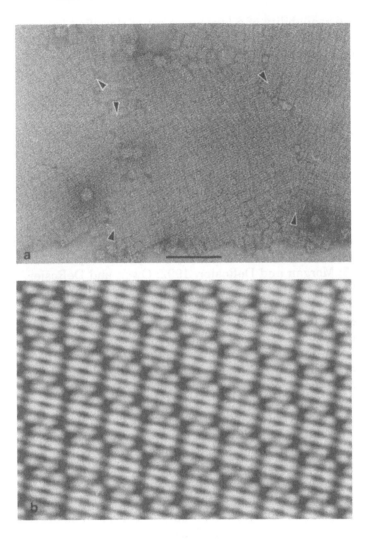

Figure 10.8: Projection image reconstruction from a 2-D crystal of the *E. coli* chaperonin GroEL.
Small 2-D crystals of the *E. coli* chaperonin GroEL, produced by the NS–CF technique (a). Note the discontinuities (arrows) and the presence of considerable disorder within the lattices. (b) A p2 crystallographic 2-D reconstruction produced within CRISP, from a well-ordered 2-D cystal of GroEL. Note the asymmetry of each GroEL 7-mer, with symmetry with the 2 × 7-mer, due to a specific 25.7° rotation at the equatorial plane (see Braig *et al.*, 1994). The scale bar (a) indicates 100 nm. Modified from Harris *et al.* (1995).

of seven cpn60 subunits with a 25.7° rotation about the central/equatorial plane (Braig *et al.*, 1994). This generates an asymmetry within the projection image of each heptameric ring, a subtle yet structurally *very important* image feature present in *Figure 10.8b*, but *not* always revealed by other published 2-D and 3-D reconstructions of the cpn60 2 × 7-mer using single particle analysis. Recovery of such information by single particle analysis requires the availability of a large data set of classified images containing several different orientations, as was the case with the KLH1 didecamer shown in *Figure 10.1*, which is derived from more than 4000 individual particles (cf. *Figure 5.45* and front cover).

10.3 Helical structures

Despite the fact that helical image processing from electron micrographs was one of the earliest areas to be successful (DeRosier and Klug, 1968; DeRosier and Moore, 1970) the widespread availability of software for performing helical processing and reconstruction of a range of different helical polymers has not emerged. This is perhaps because individuals modified the original MRC software for their own specific applications (Epstein *et al.*, 1995; Morgan and DeRosier, 1992; Owen and DeRosier, 1993; Schmidt *et al.*, 1995) rather than creating a more generally applicable and widely usable package. With the creation of the PHOELIX system for helical image processing (Carragher *et al.*, 1996), this situation may be in the process of changing.

The impressive studies of Unwin and his colleagues on tubular (helical) crystals of the acetylcholine (ACh) receptor over the past decade have resulted in the initial production of a 17 Å helical image reconstruction (Toyoshima and Unwin, 1990) and more recently a 9 Å reconstruction (Unwin, 1993, 1995) of the receptor channel. Even more elegantly, it was shown in the higher resolution study that brief (5 msec) interaction of the ACh receptor tubules with ACh immediately prior to rapid vitrification, produced an opening of the receptor channel due to a conformational change of the subunit α-helices lining the membrane-spanning pore, as shown in *Figure 10.9*. The distinct conformations of the two α-subunits of the channel complex are believed to be related to ACh-binding and channel opening (Unwin, 1996). This, the first demonstration of the structural basis for the very rapid response of the receptor channel to neurotransmitter, thereby creating a brief increase in ion permeability in the postsynaptic membrane, may be of widespread significance in neurobiology.

The way forward for macromolecular high resolution transmission electron microscopy is undoubtedly *via* image processing and 3-D reconstruction. The comments provided by some of the main exponents in this field (DeRosier, 1993; Henderson, 1995) lead one to predict steady if slow future progress towards higher resolution reconstructions from TEM images.

It remains to be said that both negative staining and cryoelectron microscopy can produce visually beautiful images of biological samples (*pretty pictures*); these can readily be compared with images of a less satisfactory nature from damaged or poorly stained material. Usually, the initially beautiful images may yield the *best* results when image processing techniques are applied, but these too have to be strictly assessed by the acuity of the human eye and accumulated experience of the human brain, when placed alongside the *raw data* in the original electron optical images. The above comments do not strictly apply to low dose electron crystallographic studies, where the low image contrast may preclude direct definition of the individual molecules or particles. Electron diffraction patterns from unstained 2-D crystals do themselves possess an inher-

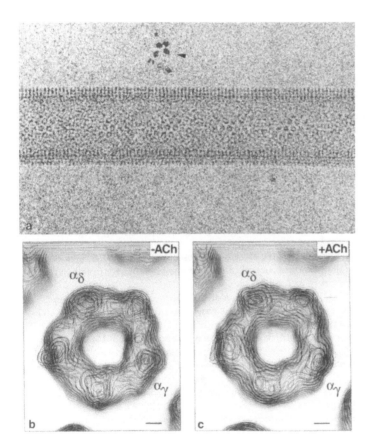

Figure 10.9: A *Torpedo* ray postsynaptic helical membrane 'tubule' with ACh receptors visible.
(a) A *Torpedo* ray postsynaptic membrane 'tubule' after spray-mixing and rapid freezing, imaged at −178°C. Crystalline arrays of ACh receptors are visible on the tubule surface, contrasted darkly against the surrounding lipid and ice. The extracellular domains of the receptors face outwards, giving the edge of the tubule a striated appearance. The presence of ferritin particles (arrow) in the vicinity of the tubules indicates that the receptors area is near the edge of a spreading spray-droplet, and that the receptors have therefore been exposed to ACh. The diameter of the tubule is 760 Å. (b and c) The extracellular domain of the ACh receptor. Views from the synaptic cleft of the mouth of the receptor channel before (b) and after (c) activation by ACh, made by stacking successive 2 Å-spaced sections on top of one another. The scale bars (b, c) indicate 10 Å. Micrograph and reconstructions courtesy of Nigel Unwin. Reprinted with permission from Unwin (1995) Acetylcholine receptor channel imaged in the open state. *Nature* 373, 37–43. Copyright 1995 Macmillan Magazines Ltd.

ent ordered beauty, as do the molecular and viral image reconstructions from many studies, particularly so with the increasingly widespread use of impressive multi-coloured reproductions in the scientific journals.

References

Braig K, Ptwinowski Z, Hegde R, Boisvert DC, Joachimiak A, Horwich AL, Sigler PB. (1994) The crystal structure of the bacterial chaperonin GroEL at 2.8 Å. *Nature* **371**, 578–586.

Carragher B, Smith PR. (1996) Advances in computational image processing for microscopy. *J. Struct. Biol.* **116**, 2–8.

Carragher B, Whittaker M, Milligan RA. (1996) Helical processing using PHOELIX. *J. Struct. Biol.* **116**, 107–112.

Chen S, Roseman AM, Hunter AS, Wood SP, Burson SG, Ranson NA, Clarke AR, Saibil HR. (1994) Location of a folding protein and shape changes in GroEL–GroES complexes imaged by cryo-electron microscopy. *Nature* **371**, 261–264.

Crowther RA. (1971) Procedures for three-dimensional reconstruction of spherical viruses by Fourier synthesis from electron micrographs. *Phil. Trans. R. Soc. Lond.* **B 261**, 221–230.

Crowther RA, Amos L. (1971) Harmonic analysis of electron microscope images with rotational symmetry. *J. Mol. Biol.* **60**, 123–130.

Crowther RA, Henderson R, Smith JM. (1996) MRC image processing programs. *J. Struct. Biol.* **116**, 9–29.

DeRosier DJ. (1993) Turn-of-the-century electron microscopy. *Curr. Biol.* **3**, 690–692.

DeRosier DJ, Klug A. (1968) Reconstruction of three-dimensional structures from electron micrographs. *Nature* **217**, 130–134.

DeRosier DJ, Moore PB. (1970) Reconstruction of three-dimensional images from electron micrographs of structures with helical symmetry. *J. Mol. Biol.* **52**, 355–369.

Epstein EF, Lu GY, Deitiker PR, Oritz I, Schmidt MF. (1995) Preliminary three-dimensional model for nematode thick filament core. *J. Struct. Biol.* **115**, 163–174.

Erickson H, Klug, A. (1971) Measurement and compensation of defocusing and aberrations by Fourier processing of electron micrographs. *Phil. Trans. R. Soc. Lond.* **B 261**, 105–118.

Harris JR, Holzenburg A. (1989) Transmission electron microscopy of negatively stained human erythrocyte catalase. *Micron Microsc. Acta* **20**, 223–238.

Harris JR, Holzenburg A. (1995) Human erythrocyte catalase: 2-D crystal nucleation and multiple 2-D crystal forms. *J. Struct. Biol.* **115**, 102–112.

Harris JR, Cejka Z, Wegener-Strake A, Gebauer W, Markl J. (1992) Two-dimensional crystallization, transmission electron microscopy and image processing of keyhole limpet haemocyanin. *Micron Microsc. Acta* **23**, 287–301.

Harris JR, Volker S, Engelhardt H, Holzenburg A. (1993) Human erythrocyte catalase: new 2-D crystal forms and image processing. *J. Struct. Biol.* **111**, 22–33.

Harris JR, Zahn R, Plückthun A. (1995) Electron microscopy of the GroEL–GroES filament. *J. Struct. Biol.* **115**, 68–77.

van Heel M, Winkler H, Orlova E, Schatz M. (1992) Structure analysis of ice-embedded particles. *Scann. Microsc.* (Suppl.) **6**, 23–42.

Henderson R. (1995) The potential and limitations of neutrons, electrons and X-rays for atomic resolution microscopy of unstained biological molecules. *Quart. Rev. Biophys.* **28**, 171–193.

Hoenger A, Sablin EP, Vale RD, Fletterick RJ, Milligan RA. (1995) Three-dimensional structure of tubulin–motor-protein complex. *Nature* **376**, 271–274.

Hovmöller S. (1992) CRISP: Crystallographic image processing on a personal computer. *Ultramicroscopy* **41**, 121–135.

Kocsis E, Cerritelli ME, Trus BL, Cheng N, Steven AC. (1995) Improved methods for determination of rotational symmetries of macromolecules. *Ultramicroscopy* **60**, 219–228.

Lamy J, Gielens C, Lambert O, Taveau JC, Motta G, Loncke P, De Geest N, Preaaux G, Lamy J. (1993) Further approaches to the quaternary structure of *Octopus* hemocyanin: a model based on immunoelectron microscopy and image processing. *Arch. Biochem. Biophys.* **305**, 17–29.

Langer T, Pfeifer G, Martin J, Baumeister W, Hartl F-U. (1992) Chaperonin-mediated protein folding: GroES binds to one end of the GroEL cylinder, which accommodates the protein substrate within its central cavity. *EMBO J.* **11**, 4757–4765.

Mellema JE, Klug A. (1972) Quaternary structure of gastropod haemocyanin. *Nature* **239**, 146–150.

Morgan DG, DeRosier D. (1992) Processing images of helical structures: a new twist. *Ultramicroscopy* **46**, 262–285.

Owen C, DeRosier D. (1993) A 13 Å map of the actin-scruin filament from the Limulus acrosomal process. *J. Cell Biol.* **123**, 337–344.

Penczek P, Radermacher M, Frank, J. (1992) Three-dimensional reconstruction of single particles embedded in ice. *Ultramicroscopy* **40**, 33–53.

Saxton WO. (1996) Semper: distortion compensation, selective averaging, 3-D reconstruction, and transfer function correction in a highly programmable system. *J. Struct. Biol.* **116**, 230–240.

Schmidt MF, Jakana J, Chiu W, Matdudaira P. (1995) 7 Å projection map of frozen, hydrated acrosomal bundle from *Limulus* sperm. *J. Struct. Biol.* **115**, 209–213.

Serysheva II, Prlova EV, Chiu W, Sherman MB, Hamilton SL, van Heel M. (1995) Electron cryomicroscopy and angular reconstitution used to visualize the skeletal muscle calcium release channel. *Nature Struct. Biol.* **2**, 18–23.

Stewart PL, Burnett RM. (1995) Adenovirus structure by X-ray crystallography and electron microscopy. In *Current Topics in Microbiology and Immunology*, Vol. 199/1: *The Molecular Repertoire of Adenoviruses I* (eds W Doerfler and P Böhm). Springer-Verlag, Berlin, pp. 25–38.

Stewart PL, Burnett RM, Cyrklaf M, Fuller SD. (1991) Image reconstruction reveals the complex molecular organization of adenovirus. *Cell* **67**, 145–154.

Stewart PL, Fuller SD, Burnett RM. (1993) Difference imaging of adenovirus: bridging the resolution gap between X-ray crystallography and electron microscopy. *EMBO J.* **12**, 2589–2599.

Toyoshima C, Unwin N. (1990) Three-dimensional structure of the acetylcholine receptor by cryoelectron microscopy and helical image reconstruction. *J. Cell Biol.* **111**, 2623–2635.

Unwin N. (1993) Nicotinic acetylcholine receptor at 9 Å resolution. *J. Mol. Biol.* **229**, 1101–1124.

Unwin N. (1995) Acetylcholine receptor channel imaged in the open state. *Nature* **373**, 37–43.

Unwin N. (1996) Projection structure of the nicotinic acetylcholine receptor: distinct conformations of the α-subunits. *J. Mol. Biol.* **257**, 586–596.

Unwin PNT, Henderson R. (1975) Molecular structure determination by electron microscopy of unstained crystalline specimens. *J. Mol. Biol.* **94**, 425–440.

Wagenknecht T, Radermacher M. (1995) Three-dimensional architecture of the skeletal muscle ryanodine receptor. *FEBS Lett.* **369**, 43–46.

Weirich TE, Ramlau R, Simon A, Hovmöller S, Zou X. (1996) A crystal structure determined with 0.02 Å accuracy by electron microscopy. *Nature* **382**, 144–146.

Zahn R, Harris JR, Pfeifer G, Plückthun A, Baumeister W. (1993) Two-dimensional crystals of the molecular chaperone GroEL reveal structural plasticity. *J. Mol. Biol.* **229**, 579–584.

Zhou ZH, Prasad BV, Jakana J, Rixon FJ, Chiu W. (1994) Protein subunit structures in the herpes simplex virus A-capsid determined from 400 kV spot-scan electron cryomicroscopy. *J. Mol. Biol.* **242**, 456–469.

Zhou ZH, Hardt S, Wang B, Sherman M, Jakana J, Chiu W. (1996) CTF determination of images of ice-embedded single particles using a graphic interface. *J. Struct. Biol.* **116**, 216–222.

Appendix A
Manufacturers and suppliers of TEM ancillary equipment and consumables

Agar Scientific Ltd, 66a Cambridge Road, Stanstead, Essex CM24 8DA, UK.

BAL-TEC AG, FL-9496 Balzers, Fürstentum, Liechtenstein.

BAL-TEC Productions Inc., 984 Southford Road, US-Middlebury, CT 06762, USA.

BIO-RAD, Microscience Division, 19, Blackstone Street, Cambridge, MA 02139, USA.

BIO-RAD, Microscience Division, Bio-Rad House, Maylands Avenue, Hemel Hempstead, Hertfordshire HP2 7TD, UK.

Denton Vacuum Inc., 1259 North Church Street, Moorestown, NJ 08057, USA.

Edwards High Vacuum International, Manor Royal, Crawley, West Sussex RH10 2LW, UK.

Electron Microscopy Sciences, 321 Morris Road, P.O. Box 251, Fort Washington, PA 19034, USA.

Ernest F. Fullam Inc., 900 Albany Shaker Road, Latham, NY 12110, USA.

Gatan Inc., 6678 Owens Drive, Pleasanton, CA 94588, USA.

Gatan GmbH, Inglostädter Strasse 40, D-80807 München, Germany.

Gatan Ltd, Ash House, 17 Medicott Close, Oakley Hay, Corby NN18 9NF, UK.

Leica, Reichert Division, Hernalser Haupstrasse 219, A-1171 Vienna, Austria.

Nanoprobes Inc., 25 E. Loop Road, Ste. 124, Stony Brook, NY 11790-3350, USA.

Oxford Instruments, Old Station Way, Eynsham, Witney, Oxford OX8 1TL, UK.

Plano, W. Plannet GmbH, Ernst-Belfort-Strasse 12, D-35578 Wetzlar, Germany.

Polysciences Inc., 400 Valley Road, Warrington, PA 18976, USA.

Polysciences Europe GmbH, Postfach 1130, D-69208, Eppelheim, Germany.

TAAB Laboratories Equipment Ltd, 3, Minerva House, Calleva Industrial Park, Aldermaston, Berkshire RG7 8NA, UK.

Ted Pella Inc., The Microscopy Supply Center, P.O. Box 492477, Redding, California, 96049-2477, USA.

VG Microtech, Bellbrook Business Park, Bolton Close, Uckfield, East Sussex TN22 1QZ, UK.

Many companies also have local/national branches or agencies; not all are listed here.

Appendix A
Manufacturers and suppliers of TEM
auxiliary equipment and consumables

Appendix B
Suppliers of commercially available 2-D and 3-D image processing software

CRISP (Trimerge, Triview) and ELD
Calidris, Manhemsvägen 4, S-191 Sollentuna, Sweden.

SEMPER
Synoptics Ltd, 271 Cambridge Science Park, Milton Road, Cambridge CB4 4WE, UK.
Synoptics Ltd, 164 CJC Highway, Cohasset, MA 02025, USA.

IMAGIC-5
Imagic Science Software GmbH, Mechlenburgische Strasse 27, D-14197, Berlin, Germany.

Index

Milton Keynes UK
Ingram Content Group UK Ltd.
UKHW031149141024
449569UK00024B/954